Kubernetes Operator
开发进阶

胡 涛（Daniel Hu）编著

机械工业出版社
China Machine Press

图书在版编目（CIP）数据

Kubernetes Operator开发进阶 / 胡涛编著. — 北京：机械工业出版社，2022.10
ISBN 978-7-111-71615-0

Ⅰ. ①K… Ⅱ. ①胡… Ⅲ. ①Linux操作系统–程序设计 Ⅳ. ①TP316.85

中国版本图书馆CIP数据核字（2022）第172470号

Kubernetes 是一个由 Google 开源的容器化应用编排系统，该系统为容器化应用提供了强大的资源管理调度、服务发现、应用扩/缩容、应用滚动更新与失败回滚等功能。Kubernetes早在几年前就已经一统容器编排领域，成为容器编排的事实标准，彻底颠覆了软件的开发和运维模式。随着以Kubernetes为中心所构建的云原生应用的不断流行，人们逐渐发现通过Kubernetes原生资源与控制器来管理复杂有状态的应用变得越来越困难。后来Kubernetes开始支持自定义资源，接着又出现了Operator模式，最后Operator模式开始逐渐流行，成为复杂有状态的应用上云的事实标准。

本书详细讲解Operator开发过程中所涉及的各个知识点，从简单的Operator示例应用入手，帮助读者快速上手Operator的开发流程，接着深入分析client-go、Deployment控制器等的源码，通过一个进阶应用的开发过程进一步详细介绍Operator开发的各方面知识。通过本书的学习，读者能够轻松掌握Operator的开发技巧，深入理解Operator的底层原理等，进而在日常工作中更好地通过Operator实现各种复杂的应用治理逻辑的开发。

无论对于云原生领域的开发工程师、测试工程师、运维工程师、软件架构师、技术经理，还是对于想要深入研究Kubernetes、掌握Kubernetes Operator开发技能的大中专院校相关专业的学生，本书都极具参考价值。

Kubernetes Operator 开发进阶

出版发行：	机械工业出版社（北京市西城区百万庄大街22号 邮政编码：100037）
责任编辑：迟振春	责任校对：肖 琳 王明欣
印　刷：北京捷迅佳彩印刷有限公司	版　次：2023年1月第1版第1次印刷
开　本：188mm×260mm　1/16	印　张：14.75
书　号：ISBN 978-7-111-71615-0	定　价：89.00元

客服电话：(010) 88361066　68326294

版权所有 • 侵权必究
封底无防伪标均为盗版

推荐序 1

经过作者多年的实践积累和长期的精心准备，这本书终于和读者见面了。我有幸成为本书的首批读者，同时也是最早参与研发 Operator 技术的工程师，内心是十分激动的：我不仅见证了这项技术被应用到越来越多的现实场景中，同时也看到这部分知识通过图书等形式记录下来，造福更多的开发者。

说起 Operator 的历史，一开始它是为了解决如何在 Kubernetes 上部署有状态应用而发明的。Kubernetes 早期只能够部署无状态应用，而对于有状态应用（如 Etcd、MySQL、Kafka 等）并没有优雅的部署方案。后来我们通过 Operator 这套机制让管理有状态应用变得跟管理无状态应用一样简单。不仅如此，Operator 的核心价值在于能够扩展 Kubernetes API，这也让越来越多的工作负载得以运行在 Kubernetes 上。阿里在落地 Kubernetes 的过程中，就通过 Operator 机制将大规模服务部署成功，并诞生了 OpenKruise 项目。

我认为本书非常值得一读，主要有以下几点理由：

首先，本书内容通俗易懂，并且从开发者的角度出发，对每一个概念都加入了相应的代码实例来辅助理解，有助于初学者快速学习相关知识。

其次，本书提供了大量的实例操作，不仅有利于提高读者的开发技能，还能够帮助读者解决在实际工作中经常遇到的各种关键问题。

再次，本书讲解某些内容时直接深入解析代码实现，让读者彻底理解其中的原理。这对于有一定基础（如 client-go）的专业人士学习 Kubernetes 的各种细节和实践操作十分有利。

最后，本书系统地讲述了整个 Operator 机制的原理和生态，而不是单一地解读某个工具或者服务。我们可以看到，以 Kubernetes 为基础的整套云原生技术被越来越多的企业所采用，这里面的每一项技术（如 Operator、Helm、Kustomize 等）都不是孤立无关的。串联好一整套云原生技术，站在新技术变革的肩膀上，才能让它们发挥出不可预估的价值！

Operator 机制创始作者之一、CNCF 应用交付领域联席主席　邓洪超

推荐序 2

我有幸经历了几代虚拟化技术的演进和整合,见证了云原生技术的星火燎原。记得还在校园的时候,我跑去中关村的一家小公司实习,天天"啃"Xen的源代码。那一年,陈海波教授的团队发表了国内第一篇SOSP论文,做嵌套的虚拟化技术CloudVisor;同年,UC Berkeley的AMPLab在NSDI上发表了有关Mesos的论文,后来衍生出Mesosphere和其上的容器编排系统Marathon。时隔不久,Google开源了Kubernetes,并在EuroSys上公开了其内部容器管理系统Borg和Omega的设计,它们可谓Kubernetes的前身。

在近十年激烈的市场竞争中,Kubernetes脱颖而出,事实上比Mesos+Marathon更具优势。Kubernetes最终获得了更广泛的社区和用户,这与它简洁的设计、强大的可扩展性、优秀的开发者体验密不可分。而这些Kubernetes的成功要素有一个重要的基石,就是它的Operator模式。Operator扩展了Kubernetes可应用和服务的场景,同时为开发者参与和扩展这些新场景提供了有效途径。

胡涛将他在GitHub和博客上连载的Kubernetes源码分析文章扩充升级为此书,系统介绍了Operator开发的知识。从设计到源码,从实例到项目,此书是国内这方面少有的原理与实践相结合的佳作,相信它能帮助国内开发者一探Kubernetes的精髓。学习和掌握Operator技术,既可以提升对云原生和容器系统的理解,又可以满足实际系统开发和运维的需要。胡涛的工作立足于Kubernetes在开发者社区中快速扩展的关键点,同时也为Kubernetes社区和生态添砖加瓦。

在思码逸,胡涛正在和我们一起打造DevOps工具链管理器DevStream,并已成功捐赠给云原生计算基金会(CNCF),继续推动云原生技术的发展。与胡涛共事,我深知他言语幽默,常常思如涌泉,期待读者能够享受他在此书中飞扬的文字。

思码逸创始人兼CEO、清华大学计算机系博士　任晶磊

前　言

我不知道你是怎么发现这本书的，不过我相信有很大的概率是你关注了我的个人微信公众号"胡说云原生"，然后认可我分享的技术文章，接着某一天从中得知我写了这本《Kubernetes Operator 开发进阶》，最后你决定支持一下这个默默写了好几年技术文章的"博主"，于是在网上下单买了这本书。假如你是我的粉丝，在这里我想对你一直以来的认可与支持表示最真诚的感谢！

在 2013 年，划时代的项目 Docker 正式发布，让容器化技术进入大众的视野。接着在 2014 年 Kubernetes 开源，2015 年 CNCF 成立，云原生浪潮就此袭来，容器编排和管理系统开始蓬勃发展，进而开始了 Kubernetes、Mesos、Swarm+Compose 三分天下的局面。

我是在 2016 年下半年开始工作的（巧的是第一个 Operator 项目也是在 2016 年下半年诞生的），也就是 CNCF 成立后一年。毕业那年我去了 H3C 的云计算部门，有幸早早接触到了 Kubernetes，然后在云原生的世界里摸爬滚打，走了无数弯路，从小白逐渐成长为一只"老鸟"，开始承担 Kubernetes 集群的维护、异常定位、优化等工作。后来凭借着在云原生领域的技术积累，我在第二份工作中开始参与整个云平台从 0 到 1 的建设。眨眼间，从毕业到现在我已经工作了 6 年，在这 6 年时间里，我的每一份工作都围绕着云原生生态技术栈。仔细想来，我在企业里落地以 Kubernetes 为核心的云原生生态技术栈，其实也是以 Kubernetes 为核心的云原生生态技术栈给了我"饭碗"。在可预见的很长的职业生涯中，我都会围绕着云原生、DevOps 等技术，想来也是和 Kubernetes 结缘颇深！

Kubernetes 一词来自希腊语，意思是"舵手"。我们知道 container 这个单词的意思是集装箱，而 Docker 的 LOGO 是一只"驮"着集装箱的鲸鱼，寓意着其"容器管理器"的定位。而 Kubernetes 以舵作为 LOGO，可见其野心是要做容器管理的"领路人"。事实上，Kubernetes 确实已经确立了"领路人"的地位，成为云原生时代的一个"云操作系统"，也成为容器编排领域的事实标准。

伴随着微服务理念的发展与流行，云上的应用逐渐变得复杂起来，在云上部署一个应用所需要的配置越来越复杂，这给云原生应用的运维带来了不小的挑战。后来 Helm 和 Kustomize 的出现分别解决了云原生应用的部署管理复杂性问题和多环境差异化配置管理问题等，在一定程度上简化了应用上云的过程。但是复杂的分布式、有状态应用除了部署复杂之外，还有大量复杂的维护工作需要站点可靠性工程师（SRE）去操心，比如数据备份、故障恢复、有条件地扩/缩容等，这些技能往往是特定领域的运维工程师擅长的，而不是一类应用所共有的属性。伴随着云上应用的复杂化，Operator 应运而生，接着站点可靠性工程师便可以通过编写 Operator 类型的运维软件来运维自己的应用，也就是将自己的领域运维经验代码化，以代码的形式管理应用，通过代码代替手动的云上运维操作来自动化地实现特定应用的运维管理。

毫无疑问，Kubernetes 的理念很酷，而 Kubernetes 里最酷的就是通过 Operator 模式实现的高度可拓展性。但是目前市面上很少有书籍介绍如何开发 Operator，以及 Operator 的底层原理等。尤其是专门讲解 Operator 的书籍，更是少之又少。我自己在学习 Operator 的过程中也是一度感觉入门很困难，走了很多弯路。很多 Operator 的原理在网上几乎找不到，要想深入学习，只能自己深入源码去摸索，而Kubernetes的源码阅读起来其实门槛不低。那时候我在网上以开源的方式连载《k8s-1.13版本源码分析》，这本电子书目前在我的GitHub（https://github.com/daniel-hutao）里标星已经接近1000了。后来我又在自己的个人博客（https://www.danielhu.cn）上连载《Kubernetes client-go 源码分析》，详细分析 Operator 开发所涉及的 client-go 底层模块的源码。所以有一天，机械工业出版社的编辑找到我，问我是否有意向出版相关图书时，我便欣然答应了。

在约稿后的大半年时间里，我几乎没有周末和假期，在本就很忙碌的本职工作之外花费了大量精力来编写本书。但是毕竟我个人的能力和精力有限，而 Kubernetes 又是一个庞大的"工程"，所以我再认真也难免会出现一些纰漏。如果读者朋友发现了书中的不妥之处，欢迎不吝指正，可以通过评论的方式将其留到本书对应的博客（https://www.danielhu.cn/advanced-kubernetes-operator/）的评论区，也可以通过我的博客或者 GitHub 等平台反馈给我。让我们互相学习，一起玩转 Operator、玩转云原生！

思码逸 DevOps 专家　胡涛

目　　录

推荐序 1
推荐序 2
前言

第一篇　入　　门

第 1 章　了解 Kubernetes ... 2
1.1　初识 Kubernetes ... 2
1.2　Kubernetes 集群的部署 ... 3
　　1.2.1　Docker 的安装 ... 4
　　1.2.2　Kind 工具介绍 ... 6
　　1.2.3　使用 Kind 快速搭建 Kubernetes 环境 ... 6
　　1.2.4　使用 Kind 搭建多节点 Kubernetes 集群环境 ... 7
　　1.2.5　Kind 用法进阶 ... 10
1.3　Kubernetes 集群的基本操作 ... 12
　　1.3.1　示例项目介绍 ... 12
　　1.3.2　基础操作演示 ... 13
　　1.3.3　小结 ... 18
1.4　Kubernetes 的核心概念 ... 18
　　1.4.1　节点 ... 18
　　1.4.2　命名空间 ... 19
　　1.4.3　容器组 ... 21
　　1.4.4　副本集 ... 22
　　1.4.5　部署 ... 23
　　1.4.6　服务 ... 24
1.5　Kubernetes 的发展历史 ... 26
1.6　本章小结 ... 27

第 2 章　开始 Operator 开发 ... 28
2.1　理解控制器模式 ... 28
　　2.1.1　生活中的控制器 ... 28
　　2.1.2　Kubernetes 中的控制器 ... 29
2.2　理解 Operator 模式 ... 30
2.3　Operator 开发环境准备 ... 31

2.4	Kubebuilder 的安装配置	31
2.5	从 Application Operator Demo 开始	32
	2.5.1 创建项目	33
	2.5.2 添加 API	35
	2.5.3 CRD 实现	38
	2.5.4 CRD 部署	39
	2.5.5 CR 部署	40
	2.5.6 Controller 实现	41
	2.5.7 启动 Controller	42
	2.5.8 部署 Controller	44
	2.5.9 资源清理	46
2.6	Operator 的发展历史	46
	2.6.1 Operator 概念的提出	46
	2.6.2 第一个 Operator 程序	47
	2.6.3 Operator 的崛起	47
2.7	本章小结	48

第二篇 进　　阶

第 3 章　Kubernetes API 介绍 50

3.1	认识 Kubernetes API	50
3.2	使用 Kubernetes API	50
	3.2.1 Curl 方式访问 API	50
	3.2.2 kubectl raw 方式访问 API	53
3.3	理解 GVK：组、版本与类型	54
3.4	本章小结	54

第 4 章　理解 client-go 55

4.1	client-go 项目介绍	55
	4.1.1 client-go 的代码库	55
	4.1.2 client-go 的包结构	56
	4.1.3 client-go 的版本规则	56
	4.1.4 获取 client-go	57
4.2	client-go 使用示例	57
	4.2.1 client-go 集群内认证配置	57
	4.2.2 client-go 集群外认证配置	60
	4.2.3 client-go 操作 Deployment	63
4.3	本章小结	67

第 5 章　client-go 源码分析············68

5.1　client-go 源码概览············68
5.1.1　关于 client-go 源码版本············68
5.1.2　client-go 模块概览············69

5.2　WorkQueue 源码分析············71
5.2.1　普通队列 Queue 的实现············71
5.2.2　延时队列 DelayingQueue 的实现············74
5.2.3　限速队列 RateLimitingQueue 的实现············79
5.2.4　小结············82

5.3　DeltaFIFO 源码分析············83
5.3.1　Queue 接口与 DeltaFIFO 的实现············83
5.3.2　queueActionLocked()方法的逻辑············85
5.3.3　Pop()方法和 Replace()方法的逻辑············86

5.4　Indexer 和 ThreadSafeStore············89
5.4.1　Indexer 接口和 cache 的实现············89
5.4.2　ThreadSafeStore 的实现············91
5.4.3　各种 Index 方法的实现············94

5.5　ListerWatcher············95
5.5.1　ListWatch 对象的初始化············95
5.5.2　ListerWatcher 接口············97
5.5.3　List-Watch 与 HTTP chunked············98

5.6　Reflector············102
5.6.1　Reflector 的启动过程············102
5.6.2　核心方法：Reflector.ListAndWatch()············102
5.6.3　核心方法：Reflector.watchHandler()············106
5.6.4　Reflector 的初始化············108
5.6.5　小结············108

5.7　Informer············109
5.7.1　Informer 就是 Controller············109
5.7.2　SharedIndexInformer 对象············113
5.7.3　sharedProcessor 对象············116
5.7.4　关于 SharedInformerFactory············119
5.7.5　小结············121

5.8　本章小结············122

第 6 章　项目核心依赖包分析············123

6.1　API 项目············123
6.2　apimachinery 项目············124
6.3　controller-runtime 项目············125
6.4　本章小结············126

第 7 章 Operator 开发进阶 127

- 7.1 进阶项目设计 127
- 7.2 准备 application-operator 项目 127
  - 7.2.1 创建新项目 127
  - 7.2.2 项目基础结构分析 128
- 7.3 定义 Application 资源 132
  - 7.3.1 添加新 API 132
  - 7.3.2 自定义新 API 133
- 7.4 实现 Application Controller 134
  - 7.4.1 实现主调谐流程 134
  - 7.4.2 实现 Deployment 调谐流程 137
  - 7.4.3 实现 Service 调谐流程 140
  - 7.4.4 设置 RBAC 权限 142
  - 7.4.5 过滤调谐事件 146
  - 7.4.6 资源别名 150
- 7.5 使用 Webhook 151
  - 7.5.1 Kubernetes API 访问控制 151
  - 7.5.2 Admission Webhook 介绍 152
  - 7.5.3 Admission Webhook 的实现 152
  - 7.5.4 cert-manager 部署 154
  - 7.5.5 Webhook 部署运行 155
  - 7.5.6 Webhook 测试 157
- 7.6 API 多版本支持 159
  - 7.6.1 实现 V2 版本 API 159
  - 7.6.2 多版本 API 部署测试 160
- 7.7 API 分组支持 163
- 7.8 本章小结 164

第 8 章 Deployment Controller 源码分析 165

- 8.1 Deployment 功能分析 165
  - 8.1.1 Deployment 基础知识 165
  - 8.1.2 Deployment 的滚动更新和回滚 167
  - 8.1.3 Deployment 的其他特性 170
  - 8.1.4 小结 171
- 8.2 Deployment 源码分析 171
  - 8.2.1 逻辑入口：startDeploymentController 171
  - 8.2.2 DeploymentController 对象初始化 172
  - 8.2.3 ResourceEventHandler 逻辑 173
  - 8.2.4 DeploymentController 的启动过程 176
- 8.3 本章小结 179

第三篇　工　　具

第 9 章　使用 Kustomize 管理配置 · 182
9.1　Kustomize 的基本概念 · 182
9.2　Kustomize 的安装 · 184
9.3　使用 Kustomize 生成资源 · 185
9.3.1　ConfigMap 生成器 · 185
9.3.2　Secret 生成器 · 188
9.3.3　使用 generatorOptions 改变默认行为 · 191
9.4　使用 Kustomize 管理公共配置项 · 192
9.5　使用 Kustomize 组合资源 · 194
9.5.1　多个资源的组合 · 194
9.5.2　给资源配置打补丁 · 196
9.6　Base 和 Overlay · 202
9.7　本章小结 · 205

第 10 章　使用 Helm 打包应用 · 206
10.1　Helm 的安装 · 206
10.2　Helm 的基本概念 · 207
10.3　Helm 的常用操作 · 208
10.3.1　搜索 Chart 包 · 208
10.3.2　安装 Chart 包 · 211
10.3.3　自定义 Chart 配置 · 213
10.3.4　Release 升级与回滚 · 216
10.3.5　Release 卸载 · 219
10.3.6　Helm 命令的常用参数 · 220
10.4　封装自己的 Chart 包 · 221
10.4.1　Chart 的目录结构 · 221
10.4.2　Chart.yaml 文件 · 222
10.4.3　Chart 依赖管理 · 223
10.5　本章小结 · 224

Kubernetes Operator 开发进阶

第一篇

入门

第 1 章
了解 Kubernetes

本章将介绍什么是 Kubernetes，为什么 Kubernetes 那么流行，Kubernetes 的发展历史和基本概念，以及如何搭建 Kubernetes 环境等。通过本章的学习，我们将对 Kubernetes 有一个整体的认识，为后面进一步学习 Kubernetes 中的 Operator 开发打下基础。

当然，有一定 Kubernetes 基础的读者可以选择直接跳过本章，从下一章内容开始学习。

1.1 初识 Kubernetes

Kubernetes 是什么？

这是我们第一次听到 Kubernetes 这个词时都会问的一个问题。官网对 Kubernetes 的介绍是：Kubernetes 是一个移植性和拓展性良好的容器化工作负载和服务管理平台，极大地简化了声明式配置和自动化运维工作。

在聊 Kubernetes 是什么之前，首先需要理解什么是容器化。本书默认读者已经熟悉容器化技术和其价值，有一定的 Docker 使用经验，所以这里不准备赘述容器化的概念。

Docker 的出现诞生了一种应用交付的新形态：通过容器化技术打包一个应用及其相关依赖，从而真正实现"一次构建到处部署"。从此运维人员基本不再需要操心基础环境不一致引入的复杂"运维矩阵"问题，一个容器镜像可以轻松地在本地开发环境、测试环境、线上生产环境或者公有云环境中几乎无差别地运行起来。

有了 Docker 之后，随之而来的问题是，一个真正在生产环境中运行的系统往往是由多个容器组成的，部署的环境也往往有多台主机，如何判断哪些容器应该运行在哪些主机上，如何快速在多台主机上启动多个容器？举一个例子来直观地感受一下这个复杂度：现在你有 100 台机器和 1000 个容器，通过 docker run 去运维将会有什么样的体验？很明显这将是一个灾难。Kubernetes 的出现就很好地解决了此类问题。

通过 Kubernetes 可以轻松地管理分布式容器化应用，轻松实现弹性伸缩、负载均衡、滚动更新、故障自愈等能力。也就是说，Kubernetes 是一个"容器编排管理解决方案"。

大家肯定记得Docker的徽标（LOGO），如图1-1所示。

图 1-1　Docker 的徽标

可以看到一只鲸鱼驮着很多的集装箱。没错，这里的集装箱也就是Container，就是我们所说的容器。Docker的徽标含义就是"容器管理工具"，其实很容易联想到现实世界中"驮"着集装箱的是大型货轮。再来看Kubernetes的徽标，如图1-2所示。

图 1-2　Kubernetes 的徽标

这是一个舵。其实Kubernetes这个单词的中文也是"舵手"的意思，来自于希腊语。可见Kubernetes从诞生之初就定位于"掌舵大货轮"，也就是成为"集装箱管理者"，或者更通俗地说，Kubernetes的含义本就是"容器管理平台"，也就是常说的"容器编排管理平台"。

1.2　Kubernetes 集群的部署

"Kubernetes集群部署"严格来说是一项复杂的技术活，有很多的可选方案。要交付一套靠拢"最佳实践"的高可用集群，有很多需要考虑的技术细节。关于如何部署"真正高可用的Kubernetes集群"不在本书的讨论范围，本节我们的目标是使用一种简单的方式快速部署一个可用的Kubernetes环境。这套环境主要也是为了满足后面的开发测试需求。

快速部署Kubernetes有好几种可选方案，比如Minikube和Kind。Minikube最先基于虚拟化实现（新版本也支持容器化实现这个方案），也就是通过VirtualBox或者KVM等在本地创建若干虚拟机，然后在虚拟机中运行Kubernetes集群，一个节点也就对应一台虚拟机。Kind通过容器化实现，也就是通过Docker在本地启动若干容器，每个容器充当Kubernetes的一个节点，在容器内再运行容器化应用。本书选择用Kind这种"容器中跑容器"的方式来搭建Kubernetes环境。当然如果大家有其他擅长的工具，也完全可以使用，我们的目的仅仅是快速部署一套可用的Kubernetes集群环境。

本书以macOS作为开发环境，使用Linux或者Windows系统作为开发环境的读者可以参考本书的方法，相应地做一些灵活调整。

下面我们将介绍如何快速搭建Kubernetes环境，有了这套环境之后就可以开始学习后面的1.3节。1.2.4节和1.2.5节可以在学习完第1章，开始准备Operator开发环境时再回过头来学习。

1.2.1 Docker 的安装

在Linux下安装Docker是一件非常简单的事情，Docker的核心原理就是基于Linux的Namespace和Cgroup等机制。不过在macOS和Windows下需要通过虚拟化技术间接使用Docker。当然，我们现在已经不需要先安装虚拟化软件，然后自己安装Linux虚拟机，再使用Docker了，而是可以直接在官网docker.com下载Docker Desktop来运行Docker程序。

我们可以在https://www.docker.com/products/docker-desktop上寻找合适的Docker Desktop版本，主要是看清楚CPU架构是Intel Chip还是Apple Chip，前者对应AMD64架构版本Mac，后者对应ARM架构的M1芯片版本Mac。下载页面大致如图1-3所示。

图 1-3　Docker Desktop 下载页面

下载完成后双击Docker.dmg文件，可以看到如图1-4所示的安装页面。

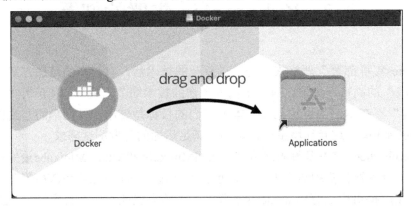

图 1-4　Docker 安装页面

接下来把Docker图标拖曳到Applications中，稍等一会儿，就可以在"启动台"中看到Docker图标，如图1-5所示。然后单击"启动台"中的Docker以打开Docker Desktop。

稍等几秒，就可以看到如图1-6所示的启动页面了。

图 1-5　Docker 的应用图标

图 1-6　Docker 的启动页面

可以单击右上角的"齿轮"按钮来修改Docker Desktop的一些配置，比如调整Docker可以使用的资源等。如果后面需要启动的容器较多导致内存不够用了，可以回到这里进行调整，比如这里把Memory调整到4.00 GB，如图1-7所示。

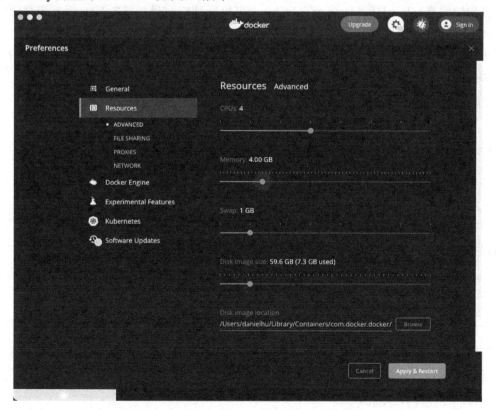

图 1-7　Docker 配置页面

修改后单击右下角的Apply & Restart按钮才会生效。

1.2.2 Kind 工具介绍

Kind（Kubernetes-in-docker）是一个使用Docker容器作为"节点"实现部署Kubernetes集群环境的工具。Kind工具主要用于Kubernetes本身的测试，目前很多需要部署到Kubernetes环境测试的项目在CI流程中都会选择用Kind来快速拉起一个Kubernetes环境，然后运行相关测试用例。

Kind本身很简单，只包含一个简单的命令行工具kind以及一个用来启动Kubernetes和systemd等的Docker镜像。我们可以这样理解Kind的原理：它通过主机上的Docker，使用封装了Kubernetes等工具的容器镜像启动一个容器，这个容器里运行了systemd，容器里的systemd可以进一步运行Docker和Kubelet等Kubernetes节点所需的基础进程，然后这些进程就可以运行kube-apiserver、kube-controller-manager、kube-scheduler、kube-proxy、CoreDNS等集群所需的组件，于是这样一个容器就组成了一个Kubernetes集群的"节点"。由此可见，Kind可以通过一个容器运行"单节点Kubernetes集群"，也可以进一步通过运行三个或更多容器实现在一台主机上运行一个"多节点Kubernetes集群"。

1.2.3 使用 Kind 快速搭建 Kubernetes 环境

现在我们来搭建Kind开发环境，在GitHub上可以看到Kind最新的Release版本和对应的Node镜像：https://github.com/kubernetes-sigs/kind/releases。

既可以选择编译好的版本，也可以直接通过go get命令来下载并编译Kind。尽量选择较新的版本，然后通过下面的命令下载安装（记得改成需要的版本号）：

```
# 方式一：选择编译好的可执行文件
cd /tmp
curl -Lo ./kind https://github.com/kubernetes-sigs/kind/releases/download/v0.12.0/kind-darwin-arm64
chmod +x ./kind
sudo mv kind /usr/local/bin/

# 方式二：通过 go get 来下载和编译
go get sigs.k8s.io/kind@v0.12.1
```

可以提前下载所需的镜像，这里选择使用1.22版本的Kubernetes镜像：

```
kindest/node:v1.22.0@sha256:b8bda84bb3a190e6e028b1760d277454a72267a5454b57db34437c34a588d047
```

然后通过下面的命令就可以快速搭建一套Kubernetes环境：

```
kind create cluster --image=kindest/node:v1.22.0 --name=dev
```

上述命令执行后的输出大致如下：

```
Creating cluster "dev" ...
 ✓ Ensuring node image (kindest/node:v1.22.0) 🖼
 ✓ Preparing nodes 📦
 ✓ Writing configuration 📜
```

```
 ✓ Starting control-plane 🕹
 ✓ Installing CNI 🔌
 ✓ Installing StorageClass 💾
Set kubectl context to "kind-dev"
You can now use your cluster with:

kubectl cluster-info --context kind-dev

Have a question, bug, or feature request? Let us know!
https://kind.sigs.k8s.io/#community 😊
```

按照命令行输出的提示，接着需要执行kubectl cluster-info --context kind-dev来切换环境，其实当前直接执行kubectl get就可以看到新搭建的Kubernetes环境，多套集群的时候才需要这样切换。

```
# kubectl get node
NAME                 STATUS   ROLES                  AGE    VERSION
dev-control-plane    Ready    control-plane,master   7m4s   v1.22.0

# kubectl get pod -n kube-system
NAME                                         READY   STATUS    RESTARTS   AGE
coredns-78fcd69978-hch75                     1/1     Running   0          10m
coredns-78fcd69978-ztqn4                     1/1     Running   0          10m
etcd-dev-control-plane                       1/1     Running   0          10m
kindnet-l8qxq                                1/1     Running   0          10m
kube-apiserver-dev-control-plane             1/1     Running   0          10m
kube-controller-manager-dev-control-plane    1/1     Running   0          10m
kube-proxy-mzfgc                             1/1     Running   0          10m
kube-scheduler-dev-control-plane             1/1     Running   0          10m
```

这样我们就快速收获了一个可以用来测试或者学习Kubernetes的环境。

1.2.4　使用 Kind 搭建多节点 Kubernetes 集群环境

最小的Kubernetes HA集群需要有3个Master节点，当然我们也可以把1个节点的all-in-one环境称为"单节点集群"。本小节接着看一下如何通过Kind来快速搭建多节点的Kubernetes集群环境。

1. Kind集群配置文件

搭建Kind环境时可以自定义配置，通过--config来指定自定义配置文件路径。Kind支持的配置格式如下：

```
# this config file contains all config fields with comments
# NOTE: this is not a particularly useful config file
kind: Cluster
apiVersion: kind.x-k8s.io/v1alpha4
# patch the generated kubeadm config with some extra settings
kubeadmConfigPatches:
- |
  apiVersion: kubelet.config.k8s.io/v1beta1
  kind: KubeletConfiguration
  evictionHard:
```

```
      nodefs.available: "0%"
# patch it further using a JSON 6902 patch
kubeadmConfigPatchesJSON6902:
- group: kubeadm.k8s.io
  version: v1beta2
  kind: ClusterConfiguration
  patch: |
    - op: add
      path: /apiServer/certSANs/-
      value: my-hostname
# 1 control plane node and 3 workers
nodes:
# the control plane node config
- role: control-plane
# the three workers
- role: worker
- role: worker
- role: worker
```

可以看到这里的配置项分为两个部分，上面是与Kubeadm如何配置Kubernetes相关的配置项，下面是与nodes角色及规模相关的配置项。不难猜到，我们需要部署多个节点的Kubernetes集群，可以通过指定nodes部分配置的方式来实现。

2. 一主三从集群搭建

我们准备一份对应的配置文件，命名为multi-node-config.yaml，内容如下：

```
kind: Cluster
apiVersion: kind.x-k8s.io/v1alpha4
nodes:
- role: control-plane
- role: worker
- role: worker
- role: worker
```

接着执行如下命令启动集群：

```
# kind create cluster --config multi-node-config.yaml \
 --image=kindest/node:v1.22.0 --name=dev4
```

等待命令执行完毕，我们可以看到类似前面单节点环境搭建时看到的输出结果，主要区别是执行步骤中多了一个Joining worker nodes：

```
Creating cluster "dev4" ...
 ✓ Ensuring node image (kindest/node:v1.22.0) 🖼
 ✓ Preparing nodes 📦 📦 📦 📦
 ✓ Writing configuration 📜
 ✓ Starting control-plane 🕹️
 ✓ Installing CNI 🔌
 ✓ Installing StorageClass 💾
 ✓ Joining worker nodes 🚜

Set kubectl context to "kind-dev4"
```

```
You can now use your cluster with:

kubectl cluster-info --context kind-dev4

Thanks for using kind! 😊
```

可以通过如下命令查看新创建的集群：

```
# kubectl cluster-info --context kind-dev4
Kubernetes control plane is running at https://127.0.0.1:51851
CoreDNS is running at
https://127.0.0.1:51851/api/v1/namespaces/kube-system/services/kube-dns:dns/proxy

To further debug and diagnose cluster problems, use 'kubectl cluster-info dump'.
# kubectl get node
NAME                  STATUS   ROLES                  AGE     VERSION
dev4-control-plane    Ready    control-plane,master   3m28s   v1.22.0
dev4-worker           Ready    <none>                 2m54s   v1.22.0
dev4-worker2          Ready    <none>                 2m54s   v1.22.0
dev4-worker3          Ready    <none>                 2m54s   v1.22.0
```

从上述命令的执行结果可以很清晰地看到这个dev4集群有1个Master节点和3个Worker节点。

3. 三主三从HA集群搭建

当然这里的HA只是表示Master节点组件会运行3个副本，一定程度上实现了Master节点没有单点故障，并不是严格意义上的"高可用"。

同样准备一份配置文件ha-config.yaml，内容如下：

```
kind: Cluster
apiVersion: kind.x-k8s.io/v1alpha4
nodes:
- role: control-plane
- role: control-plane
- role: control-plane
- role: worker
- role: worker
- role: worker
```

接着执行如下命令启动集群：

```
# kind create cluster --config ha-config.yaml \
 --image=kindest/node:v1.22.0 --name=dev6
```

等待命令执行完毕，我们可以看到熟悉的日志输出结果，和上面稍有不同，这里主要是多了Configuring the external load balancer和Joining more control-plane nodes：

```
Creating cluster "dev6" ...
 ✓ Ensuring node image (kindest/node:v1.22.0) 🖼
 ✓ Preparing nodes 📦 📦 📦 📦 📦 📦
 ✓ Configuring the external load balancer ⚖
 ✓ Writing configuration 📜
 ✓ Starting control-plane 🕹
 ✓ Installing CNI 🔌
 ✓ Installing StorageClass 💾
```

```
 ✓ Joining more control-plane nodes 🎮
 ✓ Joining worker nodes 🚜
Set kubectl context to "kind-dev6"
You can now use your cluster with:

kubectl cluster-info --context kind-dev6

Have a nice day! 👋
```

这里还可以看到几个很有趣的细节，比如Preparing nodes步骤后面的小盒子数量和节点数是相等的。另外，最后一句问候语也不是固定的。比如前面是"Thanks for using kind! 😊"，这里变成了"Have a nice day! 👋"，可见Kind背后的开发者是一群"可爱"又"有趣"的工程师。

同样，我们通过几个命令看一下刚才创建的集群：

```
# kubectl cluster-info --context kind-dev6
Kubernetes control plane is running at https://127.0.0.1:52937
CoreDNS is running at https://127.0.0.1:52937/api/v1/namespaces/kube-system/services/kube-dns:dns/proxy

To further debug and diagnose cluster problems, use 'kubectl cluster-info dump'.
# kubectl get node
NAME                   STATUS   ROLES                  AGE     VERSION
dev6-control-plane     Ready    control-plane,master   8m19s   v1.22.0
dev6-control-plane2    Ready    control-plane,master   7m46s   v1.22.0
dev6-control-plane3    Ready    control-plane,master   7m20s   v1.22.0
dev6-worker            Ready    <none>                 7m      v1.22.0
dev6-worker2           Ready    <none>                 7m      v1.22.0
dev6-worker3           Ready    <none>                 7m      v1.22.0
```

从上述命令的执行结果可以很清晰地看到这个dev6集群有3个Master节点和3个Worker节点。至此，我们已经掌握了如何通过Kind来非常轻松地搭建多节点的Kubernetes集群环境，后面大家可以根据自己的需要来选择节点规模和角色，以搭建合适的测试环境。

1.2.5　Kind 用法进阶

通过前面几节的学习，我们已经掌握了使用Kind搭建各种类型的集群。但是要用好这些集群，还需要掌握一些运维技巧。本小节学习Kind集群的一些进阶操作。

1. 端口映射

设想一种场景：在Kind集群中运行一个Nginx容器服务，监听80端口对外暴露，这时在另一台机器上能不能访问Kind集群所在机器的80端口，进而访问这个Nginx服务呢？其实不行，这两个80端口明显不在一个网络命名空间。我们可以通过如下方式来配置端口映射，从而解决这类问题。

在配置文件中增加extraPortMappings配置项：

```
kind: Cluster
apiVersion: kind.x-k8s.io/v1alpha4
nodes:
```

```
- role: control-plane
  extraPortMappings:
  - containerPort: 80
    hostPort: 80
    listenAddress: "0.0.0.0"
    protocol: tcp
```

这样，搭建出来的Kubernetes集群中使用NodePort暴露80端口或者使用hostNetwork方式暴露80端口的Pod就可以通过主机的80端口来访问了。

2. 暴露kube-apiserver

有时我们会在一台计算机上使用Kind搭建一套Kubernetes环境，在另一台机器上编写代码，这时会发现我们无法连接到Kind集群中的kube-apiserver来调试Operator程序。其实这是因为默认配置下kube-apiserver监听127.0.0.1和随机端口，要从外部访问就需要把kube-apiserver监听的网卡改成非lo（代表127.0.0.1，即localhost）的对外网卡，比如eth0。

同样，我们通过配置文件自定义来实现这一需求，添加networking.apiServerAddress配置项，值是本地网卡IP（可根据实际情况修改）：

```
kind: Cluster
apiVersion: kind.x-k8s.io/v1alpha4
networking:
  apiServerAddress: "192.168.39.1"
```

3. 启用Feature Gates

如果要使用一些Alpha阶段的特性，就需要通过配置Feature Gates来实现。在使用Kubeadm搭建环境时，可以通过配置ClusterConfiguration来实现这个需求，Kubeadm被Kind封装后，在Kind中如何启用Feature Gates呢？

方案如下（FeatureGateName就是需要启用的Feature Gates名字）：

```
kind: Cluster
apiVersion: kind.x-k8s.io/v1alpha4
featureGates:
  FeatureGateName: true
```

4. 导入镜像

通过Kind搭建的环境本质是运行在一个容器内，宿主机上的镜像默认不能被Kind环境所识别，这时可以通过如下方式导入镜像：

```
# 假如需要的镜像是my-operator:v1
kind load docker-image my-operator:v1 --name dev
# 假如需要的镜像是一个tar包my-operator.tar
kind load image-archive my-operator.tar --name dev
```

知道了这个方法后，如果需要构建一个新镜像放到Kind环境中运行，就可以通过类似如下步骤来实现：

```
docker build -t my-operator:v1 ./my-operator-dir
kind load docker-image my-operator:v1
```

```
kubectl apply -f my-operator.yaml
```

怎么查看当前Kind环境中有哪些镜像呢？也很简单，可以这样：

```
docker exec -it dev-control-plane crictl images
```

其中dev-control-plane是Node容器名，有多套环境时这个名字需要灵活切换。另外，可以通过crictl -h查看crictl所支持的其他命令，比如crictl rmi <image_name>可以用于删除镜像等。

1.3 Kubernetes 集群的基本操作

本节将通过一个简单的应用演示来建立对Kubernetes的直观认识，这个过程有利于对基本概念的理解。本节的内容涉及一些术语，如果想提前了解，可以穿插着翻阅1.4节来辅助本节的学习。

1.3.1 示例项目介绍

我们将部署一个PHP Guestbook，这是一个两层架构的Web应用，这个示例原型来自官网：https://kubernetes.io/docs/tutorials/stateless-application/guestbook/。我们对其配置做一些简化，用于介绍一些常用的Kubernetes基本命令等。这个Guestbook架构如图1-8所示。

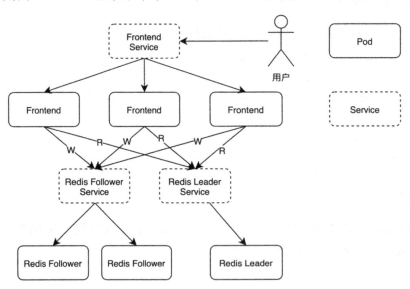

图 1-8　Guestbook 架构

图中实线圆角矩形表示的是Pod，虚线圆角矩形表示的是Service（服务），Service实现了类似F5的负载均衡能力。我们会部署3个副本的前端Pod以模拟高可用。用户请求流量会通过一个Service随机负载到某个前端Pod上，然后这个前端Pod同样通过Service访问后端Redis存储，这里会区分读操作和写操作，写操作会负载到Redis Leader上，而读操作会负载到Redis Follower上。Redis Follower也有两个副本，读操作在这里同样是利用Service实现负载均衡。

1.3.2 基础操作演示

1. 部署Redis Leader Deployment

准备配置文件redis-leader-deployment.yaml：

```yaml
apiVersion: apps/v1
kind: Deployment
metadata:
  name: redis-leader
  labels:
    app: redis
    role: leader
    tier: backend
spec:
  replicas: 1
  selector:
    matchLabels:
      app: redis
  template:
    metadata:
      labels:
        app: redis
        role: leader
        tier: backend
    spec:
      containers:
      - name: leader
        image: "docker.io/redis:6.0.5"
        ports:
        - containerPort: 6379
```

这里有一些关键信息，可以大致看一下以下配置项的含义：

- kind：声明当前资源的类型是Deployment。
- metadata：描述这个Deployment的一些元数据，主要是name和labels。
- spec：从这里开始往下都是定义当前资源的期望状态，也就是描述这个Deployment应该"怎样运行"。
- spec.replicas：声明这个Deployment需要运行多少个Pod的副本，这里是1。
- spec.template：定义Deployment用来创建Pod的模板。
- xxx.image：配置最终Pod启动时所使用的镜像。

将这个配置保存到磁盘上，然后运行如下命令创建Deployment：

```
# kubectl apply -f redis-leader-deployment.yaml
deployment.apps/redis-leader created
```

接着检查一下创建出来的资源：

```
# kubectl get deployment
NAME                    READY   UP-TO-DATE   AVAILABLE   AGE
```

```
redis-leader                              1/1       1         1         62s
# kubectl get pod
NAME                                      READY     STATUS    RESTARTS  AGE
redis-leader-856f66947-crjqs              1/1       Running   0         68s
```

2. 创建Redis Leader Service

同样准备配置文件redis-leader-service.yaml：

```
apiVersion: v1
kind: Service
metadata:
  name: redis-leader
  labels:
    app: redis
    role: leader
spec:
  ports:
  - port: 6379
    targetPort: 6379
  selector:
    app: redis
    role: leader
```

注意到这里kind变成了Service，下面的配置内容也不相同了。Service的核心配置内容是spec.ports部分，用于把redis-leader Pod的6379端口对应到redis-leader Service的6379端口，使得访问这个Service的6379端口能够间接地访问后面的Pod。另外，spec.selector和前面Pod的labels是对应的，指定了当前Service的流量转发到"哪些Pod"上。

尝试创建这个资源，然后看一下效果：

```
# kubectl apply -f redis-leader-service.yaml
service/redis-leader created
# kubectl get service
NAME            TYPE        CLUSTER-IP       EXTERNAL-IP   PORT(S)    AGE
kubernetes      ClusterIP   10.96.0.1        <none>        443/TCP    4h53m
redis-leader    ClusterIP   10.96.144.168    <none>        6379/TCP   6s
```

3. 部署Redis Follower Deployment

准备配置文件redis-follower-deployment.yaml：

```
apiVersion: apps/v1
kind: Deployment
metadata:
  name: redis-follower
  labels:
    app: redis
    role: follower
spec:
  replicas: 2
  selector:
    matchLabels:
```

```
    app: redis
  template:
    metadata:
      labels:
        app: redis
        role: follower
        tier: backend
    spec:
      containers:
      - name: follower
        image: gcr.io/google_samples/gb-redis-follower:v2
        ports:
        - containerPort: 6379
```

这里我们指定的replicas是2，也就是Redis Follower会被创建出两个Pod副本。继续尝试创建资源：

```
# kubectl apply -f redis-follower-deployment.yaml
deployment.apps/redis-follower created
# kubectl get deployment
NAME             READY   UP-TO-DATE   AVAILABLE   AGE
redis-follower   2/2     2            2           6s
redis-leader     1/1     1            1           37m
# kubectl get pod
NAME                              READY   STATUS    RESTARTS   AGE
redis-follower-67cbd9584d-8b26w   1/1     Running   0          12s
redis-follower-67cbd9584d-bln8s   1/1     Running   0          12s
redis-leader-856f66947-crjqs      1/1     Running   0          37m
```

4. 创建Redis Follower Service

准备配置文件redis-follower-service.yaml：

```
apiVersion: v1
kind: Service
metadata:
  name: redis-follower
  labels:
    app: redis
    role: follower
spec:
  ports:
  - port: 6379
  selector:
    app: redis
    role: follower
```

然后创建资源，看一下效果：

```
# kubectl apply -f redis-follower-service.yaml
service/redis-follower created
# kubectl get service
```

```
NAME             TYPE        CLUSTER-IP      EXTERNAL-IP   PORT(S)    AGE
kubernetes       ClusterIP   10.96.0.1       <none>        443/TCP    5h12m
redis-follower   ClusterIP   10.96.125.92    <none>        6379/TCP   16m
redis-leader     ClusterIP   10.96.144.168   <none>        6379/TCP   19m
```

至此，Redis就部署完成了。接下来继续部署Frontend服务。

5. 部署Frontend Deployment

准备配置文件frontend-deployment.yaml：

```yaml
apiVersion: apps/v1
kind: Deployment
metadata:
  name: frontend
spec:
  replicas: 3
  selector:
    matchLabels:
      app: guestbook
  template:
    metadata:
      labels:
        app: guestbook
    spec:
      containers:
      - name: php-redis
        image: gcr.io/google_samples/gb-frontend:v5
        env:
        - name: GET_HOSTS_FROM
          value: "dns"
        ports:
        - containerPort: 80
```

尝试创建资源，然后看一下效果：

```
# kubectl apply -f frontend-deployment.yaml
deployment.apps/frontend created
# kubectl get deployment
NAME             READY   UP-TO-DATE   AVAILABLE   AGE
frontend         3/3     3            3           35s
redis-follower   2/2     2            2           4m30s
redis-leader     1/1     1            1           41m
# kubectl get pod
NAME                              READY   STATUS    RESTARTS   AGE
frontend-77dd98787d-jkbz5         1/1     Running   0          40s
frontend-77dd98787d-nm8fw         1/1     Running   0          40s
frontend-77dd98787d-zmg2g         1/1     Running   0          40s
redis-follower-67cbd9584d-8b26w   1/1     Running   0          4m35s
redis-follower-67cbd9584d-bln8s   1/1     Running   0          4m35s
redis-leader-856f66947-crjqs      1/1     Running   0          41m
```

6. 创建Frontend Service

同样准备配置文件frontend-service.yaml：

```yaml
apiVersion: v1
kind: Service
metadata:
  name: frontend
  labels:
    app: guestbook
spec:
  ports:
  - port: 80
  selector:
    app: guestbook
```

尝试创建这个Service，然后看一下效果：

```
# kubectl apply -f frontend-service.yaml
service/frontend created
# kubectl get service
NAME             TYPE        CLUSTER-IP       EXTERNAL-IP   PORT(S)    AGE
frontend         ClusterIP   10.96.248.235    <none>        80/TCP     6s
kubernetes       ClusterIP   10.96.0.1        <none>        443/TCP    5h24m
redis-follower   ClusterIP   10.96.125.92     <none>        6379/TCP   29m
redis-leader     ClusterIP   10.96.144.168    <none>        6379/TCP   32m
```

至此，所有资源就已经创建好了。接下来看一下这个Guestbook程序是怎么使用的。

7. 查看Guestbook效果

访问页面之前，我们需要通过port-forward来暴露Frontend服务：

```
kubectl port-forward svc/frontend 8080:80
```

然后通过浏览器访问http://127.0.0.1:8080/，就可以看到如图1-9所示的页面。

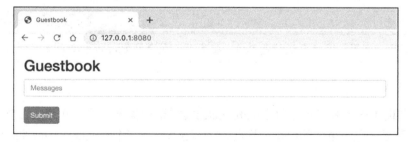

图1-9　访问页面

我们可以先试一下功能是否正常，在输入框随意输入一些内容，单击Submit按钮，如图1-10所示。

在输入框中输入了一个字符串，然后单击Submit按钮时，这个数据就被存到Redis Leader中了。接着下面显示出这个字符串，如果刷新页面，可以看到这个字符串还在，也就是从Redis Follower中读取到的数据。

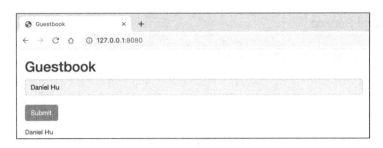

图 1-10　输入内容并提交

1.3.3　小结

本节通过Guestbook应用展示了一个两层架构的无状态容器化服务部署在Kubernetes上的流程，在这个过程中介绍了一些Kubernetes集群的基础操作，希望大家对Kubernetes操作已经有了一个感性认识。

1.4　Kubernetes的核心概念

Kubernetes引入了很多新的概念和专业术语，本节我们来梳理一下Kubernetes中的核心概念，以便大家对Kubernetes的工作原理有一个整体的了解。

1.4.1　节点

在部署了一套Kubernetes环境之后，整个Kubernetes环境就被称为一个集群（Cluster）。一个集群自然是由很多的工作节点（Node）组成的，一个工作节点对应一台主机。工作节点分为Master节点和Worker节点两种。

> **注　意**
>
> 本小节提到的"容器"准确来说就是Pod，在下面正式介绍Pod概念之前暂且叫作"容器"。

1. Master节点

Master节点也叫作控制节点，每个Kubernetes集群至少需要包含一个Master节点。如果是高可用部署，就需要三个或更多的Master节点。Master节点上运行着Kubernetes控制面的几个关键组件：

- API Server：集群所有资源的增、删、改、查入口，我们通过kubectl命令对集群进行的各种运维操作本质都是发送HTTP请求给API Server，进而触发各个组件的运行，然后相互协作，以完成一系列逻辑处理和结果响应。
- Controller Manager：集群资源"大管家"，我们在Kubernetes集群中创建的各种资源的生命周期基本上都是由Controller Manager来管理的。

- Scheduler：调度器解决的是容器调度问题，也就是如何决定某个容器运行在哪台主机之上的问题。
- ETCD：严格来说，ETCD并不是Kubernetes的一部分，也不一定和Master节点组件运行在一起。Kubernetes使用ETCD作为集群数据存储的唯一方案，也就是我们创建的Kubernetes资源都是记录在ETCD里面的。一般情况下，ETCD会和Master节点服务部署在一起，可以认为ETCD和API Server、Controller Manager、Scheduler一起组成了"控制节点"。

2. Worker节点

Kubernetes集群中除了Master之外的节点都称为Worker节点。顾名思义，Worker节点是用来具体承载容器化应用负载的。当然，这些应用会以何种方式运行在哪个Worker节点上，是由Master节点决定的。

Worker节点上的关键组件是：

- Kubelet：负责容器的创建、销毁等生命周期管理。
- Proxy：用于实现容器的服务发现和负载均衡相关能力，也就是Service的实现。后面会介绍Service的概念。
- Docker：严格来讲，Docker也不是Kubernetes的一部分。更严格来说，Docker也不是必需的，任何实现CRI规范的容器引擎都可以接入进来充当Kubernetes集群的容器运行时（runtime）。不过，在很长时间内，Kubernetes还是会与Docker一起使用，由Docker（或者说Containerd）工作在Kubelet的下层来具体实现容器的生命周期管理。

1.4.2 命名空间

Kubernetes提供了一种对资源进行分组的机制，也就是命名空间（Namespace）。我们可以通过如下命令查看当前集群中有哪些命名空间：

```
# kubectl get namespace
NAME                STATUS   AGE
default             Active   6d10h
kube-node-lease     Active   6d10h
kube-public         Active   6d10h
kube-system         Active   6d10h
local-path-storage  Active   6d10h
```

可以看到当前集群中已经有很多命名空间了。大多数资源都是区分命名空间的，比如Deployment、Service、Pod等。如果我们在创建这类资源的时候没有指定命名空间，那么它们会被默认放置到default命名空间下。当然，也有些资源是不支持区分命名空间的，比如Node、PersistentVolume、StorageClass等。目前Kubernetes API（v1.22）资源中大约有32种资源区分命名空间，26种资源不区分命名空间。可以通过如下命令来查看某个资源是否区分命名空间：

```
# kubectl api-resources --namespaced=true
NAME         SHORTNAMES   APIVERSION   NAMESPACED   KIND
bindings                  v1           true         Binding
configmaps   cm           v1           true         ConfigMap
```

```
# ...
# kubectl api-resources --namespaced=false
NAME                SHORTNAMES    APIVERSION    NAMESPACED    KIND
componentstatuses   cs            v1            false         ComponentStatus
namespaces          ns            v1            false         Namespace
# ...
```

命名空间一般被用在多租户隔离场景或者多环境隔离场景，分别设想一下这两种场景：

1）多租户隔离：假如公司现在只有3台服务器资源，搭建了一套3节点的Kubernetes集群环境，但是有多个开发团队需要在这套Kubernetes集群里测试自己的容器化应用，这个时候为了互相不干扰，我们就可以创建多个命名空间，分别叫作dev-team-1、dev-team-2……然后分别分配给不同的团队使用，从而实现多租户隔离。

2）多环境隔离：与上一个场景一样，我们只有一套3节点的Kubernetes集群环境，比如开发一个软件，在开发过程中至少需要开发环境、测试环境等。这时我们同样可以新建多个命名空间，分别叫作develop、test……然后用于不同的使用场景，从而实现多环境隔离。

在创建资源时，可以通过指定命名空间来将该资源放置到指定命名空间下，比如前面用到的redis-leader。

步骤01 创建demo命名空间：

```
# kubectl create namespace demo
namespace/demo create
```

步骤02 更新相应的YAML文件：

```
apiVersion: apps/v1
kind: Deployment
metadata:
  name: redis-leader
  namespace: demo
  labels:
    app: redis
    role: leader
spec:
  replicas: 1
// ...
```

步骤03 创建资源：

```
# kubectl apply -f redis-leader-deployment.yaml
deployment.apps/redis-leader created
```

步骤04 查询资源：

```
# kubectl get pod
No resources found in default namespace.
# kubectl get pod -n demo
NAME                          READY    STATUS     RESTARTS    AGE
redis-leader-856f66947-bxkvn  1/1      Running    0           11s
```

这里可以看到在不使用-n指定命名空间时查询的其实是默认的default命名空间。除了默认的命名空间之外，若要区分命名空间的资源查询，则需要以-n来显式地指定命名空间。

1.4.3 容器组

容器组（Pod）是Kubernetes中基础的资源，是最小的部署单元。一个容器组由一个或多个容器组成，这些容器共享存储和网络资源。另外，每个容器组中会有一个Pause容器，这个容器也就是其他业务容器的"根容器"。

容器组被设计为一个Pause容器+多个业务容器的方式。这种设计有什么好处呢？我们设想一下多个紧密相关的容器如果部署在一起组成一个容器组会面临哪些问题。

比如3个容器一起运行，这时候有1个容器挂了，那么应该怎样表示容器组的状态呢？是否需要重建这个容器组来恢复挂掉的容器？再比如3个容器启动的顺序不同,这时要实现共享网络等命名空间，应该在哪个容器的启动过程中去新建这个网络命名空间呢，又由哪个容器来坐享其成地直接加入前面一个容器创建好的命名空间呢？

引入一个Pause容器就解决了这一系列的问题，我们直接把Pause有没有挂掉当作整个容器组有没有挂掉的判断依据，直接在创建Pause容器的时候创建好需要的网络等命名空间，其他容器都加入这个命名空间里，这样就实现了这些容器可以通过lo网卡互相访问，感觉好像彼此在同一个操作系统中一样。同时，Pause容器使用很短的一段汇编代码实现，足够快速且稳定，所以也不用担心额外的性能损耗等问题。

容器组的整体结构大致如图1-11所示。

我们来准备创建一个容器组使用的配置：

```
cat <<EOF > nginx-pod.yaml
apiVersion: v1
kind: Pod
metadata:
  name: nginx
  labels:
    app: nginx
spec:
  containers:
  - name: nginx
image: nginx:1.16
EOF
```

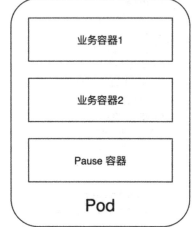

图 1-11 容器组的整体结构

然后执行如下命令：

```
# kubectl apply -f nginx-pod.yaml
pod/nginx created
```

可以看到这个容器组的状态已经处于运行中了：

```
# kubectl get pod
NAME    READY   STATUS    RESTARTS   AGE
nginx   1/1     Running   0          32s
```

1.4.4 副本集

有了容器组之后，我们很快就会面临一个问题，就是一个容器组需要运行多个副本，这时候如何管理这些副本呢？

假如需要运行一个8个副本的容器组，也就是这些副本需要同时启动，配置也完全一致，这时需要写8次YAML配置文件，然后每个配置去小心翼翼地修改一个唯一的名字，最后分别应用吗？显然不是，在容器组之上有一个管理资源叫作副本集（ReplicaSet），其作用主要是管理一定数量的容器组同时运行。

前面我们创建容器组时有这样一段配置：

```
labels:
  app: nginx
```

副本集的工作原理就是通过一个selector来指定其管理的容器组集合，并且通过一个replica配置来描述容器组的副本数量，通过一个templatc配置来描述创建容器组所用的配置模板，就像下面这样：

```
cat <<EOF > nginx-rs.yaml
apiVersion: apps/v1
kind: ReplicaSet
metadata:
  name: nginx
  labels:
    app: nginx
spec:
  replicas: 3
  selector:
    matchLabels:
      app: nginx
  template:
    metadata:
      labels:
        app: nginx
    spec:
      containers:
      - name: nginx
        image: nginx:1.16
EOF
```

接下来同样可以通过如下命令来创建这个副本集：

```
# kubectl apply -f nginx-rs.yaml
replicaset.apps/nginx created
```

执行结果如下：

```
# kubectl get rs
NAME    DESIRED   CURRENT   READY   AGE
nginx   3         3         3       6s
# kubectl get pod
```

```
NAME            READY     STATUS      RESTARTS    AGE
nginx-b2cgf     1/1       Running     0           9s
nginx-bq6sc     1/1       Running     0           9s
nginx-cb8hm     1/1       Running     0           9s
```

可以看到三个副本的nginx启动了，通过副本集实现了只修改一个副本配置就启动多个相同容器组的功能。

当然，副本集的主要作用并不是简单地帮我们去创建多个容器组，而是"确保集群中运行的容器组数和声明的一致"，也就是我们把一个副本集的副本设置为3，这时副本集背后的控制器会先检查集群中和selector匹配的容器组数量，如果小于3则创建，如果大于3则删除，如果相等则什么也不做。

但是我们一般不需要直接创建副本集资源，而是使用一个更高层的抽象资源部署。下一小节介绍资源部署。

1.4.5 部署

本书第8章会详细分析部署的功能特性和源码实现，如果大家感兴趣，也可以在看完本小节后直接跳到后面对应的章节继续学习部署的进阶知识。

前面提到部署是副本集之上的更高层抽象，我们来思考一下为什么需要这一层抽象。比如现在我们通过副本集启动了一个有3个副本的nginx服务，nginx版本用的是1.14。这时出于某种原因，需要将nginx版本更新到1.16，但是这个过程中nginx服务不能中断。怎么实现呢？

部署的出现主要就是为了解决这类问题，也就是滚动更新与回滚的能力。通过部署来管理副本集，就能够通过类似这样的操作实现滚动更新：新创建一个nginx 1.16版本的副本集，副本数设置为1，等这个容器组启动完成之后，将旧副本集的副本数减少到2，然后将新副本集的副本数设置为2……通过这样一系列的操作实现在服务不中断的情况下，完成集群中容器组的逐步升级。

创建一个部署同样很简单：

```
cat <<EOF >./nginx-deployment.yaml
apiVersion: apps/v1
kind: Deployment
metadata:
  name: nginx
  labels:
    app: nginx
spec:
  replicas: 3
  selector:
    matchLabels:
      app: nginx
  template:
    metadata:
      labels:
        app: nginx
    spec:
```

```
            containers:
            - name: nginx
              image: nginx:1.16
              ports:
              - containerPort: 80
EOF
```

同样，我们通过apply命令来创建这个资源：

```
# kubectl apply -f nginx-deployment.yaml
deployment.apps/nginx created
```

稍等一会儿就可以看到部署、副本集、容器组等资源就绪了：

```
# kubectl get deploy
NAME    READY   UP-TO-DATE   AVAILABLE   AGE
nginx   3/3     3            3           8s
# kubectl get rs
NAME               DESIRED   CURRENT   READY   AGE
nginx-644599b9c9   3         3         3       14s
# kubectl get pod
NAME                     READY   STATUS    RESTARTS   AGE
nginx-644599b9c9-6d7xs   1/1     Running   0          16s
nginx-644599b9c9-fz9qc   1/1     Running   0          14s
nginx-644599b9c9-ncw4w   1/1     Running   0          15s
```

不过我们这里只介绍部署的基础用法，不展开来讲，大家如果感兴趣，可以直接跳到后面第8章继续深入学习部署的其他功能特性的用法。

1.4.6　服务

我们学习了容器组、副本集、部署等资源对象后，已经能够很容易地创建一个多副本的无状态容器化应用，下一个问题就是：这些有着不同IP的容器组如何对外暴露服务（Service）呢？

没错，通过服务资源就可以简单地实现一个多副本应用的负载均衡，也就是我们对外暴露一个服务，调用方的流量经过服务后，会随机被转发到后端的某个容器组上。

回到上面介绍部署时使用的例子——我们创建的3个副本的nginx，容器组中有这样几行配置：

```
ports:
- containerPort: 80
```

也就是说，每个容器组其实都监听容器的80端口，我们可以通过服务来将流量转发到这个80端口：

```
cat <<EOF >./nginx-svc.yaml
apiVersion: v1
kind: Service
metadata:
  name: nginx
spec:
```

```yaml
  type: NodePort
  selector:
    app: nginx
  ports:
    - protocol: TCP
      port: 8080
      targetPort: 80
      nodePort: 30080
EOF
```

这里我们用了NodePort类型的服务，因为默认的ClusterIP方式暴露的端口在Kind方式搭建的集群中并不能直接被访问，所以这里我们将其转到节点的30080端口。

然后同样通过apply的方式创建：

```
# kubectl apply -f nginx-svc.yaml
service/nginx created
```

这时已经可以这样访问nginx了：

```
# curl localhost:30080
<!DOCTYPE html>
<html>
<head>
<title>Welcome to nginx!</title>
<style>
    body {
        width: 35em;
        margin: 0 auto;
        font-family: Tahoma, Verdana, Arial, sans-serif;
    }
</style>
</head>
<body>
<h1>Welcome to nginx!</h1>
<p>If you see this page, the nginx web server is successfully installed and
working. Further configuration is required.</p>

<p>For online documentation and support please refer to
<a href="http://nginx.org/">nginx.org</a>.<br/>
Commercial support is available at
<a href="http://nginx.com/">nginx.com</a>.</p>

<p><em>Thank you for using nginx.</em></p>
</body>
</html>
```

当然通过浏览器也是可以访问的，如图1-12所示。

这里还有一个细节问题，就是Kind集群里面的NodePort类型Service暴露的服务要在Kind集群外面访问到，需要再配置好端口映射。还记得我们在1.2.5节介绍的端口映射配置方法吗？没错，这里需要把30080端口映射出来，对应的Kind配置是这样的：

```
kind: Cluster
apiVersion: kind.x-k8s.io/v1alpha4
```

```
nodes:
- role: control-plane
  extraPortMappings:
  - containerPort: 30080
    hostPort: 30080
    listenAddress: "0.0.0.0"
    protocol: tcp
```

图 1-12　浏览器访问网页

有了这个端口映射，我们访问本地宿主机的30080端口，请求流量才会被转发到容器内的30080端口；而容器内的30080端口对于Kind集群中的Kubernetes来说，就是主机的30080端口，因此Service的NodePort从这里的30080端口接收到请求，进而转发到Cluster IP的8080端口，再随机转发到后端容器的80端口。

1.5　Kubernetes 的发展历史

如今以Kubernetes为核心的"云原生"技术生态已经彻底改变了软件交付形态，本节看一下Kubernetes的发展历史。

早在2006年，Google公司就开源了Process Container技术，也就是后来的Cgroup，在2008年，Cgroup整合进了Linux内核，同一年 LXC（Linux Container）技术也开始有了雏形。后来在2011年，Cloud Foundry开发了一个名为Warden的容器管理系统，2013年，Google公司开源了一个容器系统LMCTFY，当然这些都没有让容器化技术在业界流行起来。

真正划时代的事件是2013年Docker项目的发布，Docker让容器化技术开始进入大众视野，也让容器化技术几乎和Docker这个词等价，以至于相当一部分人会把一个容器称为"一个Docker"。

Docker的流行让大家开始把自己的传统应用逐步转向容器化构建，接着在2014年Kubernetes项目开源，2015年CNCF（Cloud Native Computing Foundation，云原生计算基金会）成立，云原生浪潮就此袭来，那几年各种容器编排和管理系统蓬勃发展，一度Kubernetes、Mesos、Swarm+Compose几种解决方案形成三足鼎立之势。不过，现在用户已经不需要纠结选择哪种容器编排管理方案了，在2017年前后，Kubernetes就已经开始与其他几种解决方案拉开距离，成为容器编排管理的事实标准。如今就更不用说了，Kubernetes已经足够强大和流行，几乎所有新开发的应用能容器化的都会容器化，容器化后稍有规模就会用上Kubernetes。

Kubernetes脱胎于Google公司内部已经运行了十多年的Borg系统。早在十几年前，Google公司就已经开始部署大规模容器化应用，Borg系统正式承载这些容器化应用，是实现海量容器化应用生命周期管理的秘密武器。Borg系统汇集了Google这么多年容器化应用的运维经验，而Kubernetes站在Borg的肩膀上，继承了Borg系统的精华，但是又没有Borg那般复杂，而是更加注重模块化、拓展性等，更加易理解。因此，Kubernetes开源后一骑绝尘，迅速流行开来。

1.6 本章小结

本章我们从Docker、Kubernetes的安装开始，介绍如何在本地快速搭建Kubernetes开发测试环境，进而通过一个示例程序演示了Kubernetes的一些基础操作，最后讲解了Kubernetes的核心概念，并在讲解每个概念的同时辅以相应的实例演示。当然，本章并不想详细地讲解Kubernetes的各种知识点，而是希望带读者入门Kubernetes，让读者能简单地使用Kubernetes，从而为后面进一步系统地学习Kubernetes打下基础。

第 2 章
开始 Operator 开发

本章将正式开始介绍Operator开发，快速讲解Operator模式是什么，并通过一个Demo程序讲解Operator，希望能够让大家对Operator的开发过程有一个整体理解；接着会介绍Operator的起源、发展历程等背景知识。

2.1 理解控制器模式

控制器模式广泛应用在机械自动化等领域，Kubernetes的核心能力就建立在"声明式API"和"控制器模式"之上。本节将从生活中的控制器模式的应用出发，介绍Kubernetes中是如何使用控制器模式的。

2.1.1 生活中的控制器

控制器模式在日常生活中的应用非常广泛。我们先来思考一下夏天空调制冷时的工作流程，如图2-1所示。

空调制冷过程如下：

1）启动后设置一个温度，比如25℃。
2）空调检测室温，确定室温是否高于25℃。
3）如果高于25℃，就开始制冷，过一会儿再监测室温变化。
4）如果不高于25℃，就保持静默，直到室温超过25℃。

这其实就是控制器模式的典型工作流程，外部输入一个"期望值"，一个"控制器"不断观测"环境状态"的"实际值"和"期望值"之间的差异，然后不断调整"实际值"向"期望值"靠拢。这个过程可以称为"调谐"。

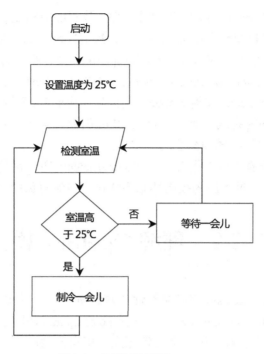

图 2-1　空调制冷流程

2.1.2　Kubernetes 中的控制器

Kubernetes中通过"声明式API"定义了一系列的"资源对象",然后通过很多"控制器"来"调谐"这些资源对象的实际状态向期望状态靠拢,从而实现整个集群"尽可能"靠拢配置中声明的期望状态。

我们知道在Kubernetes的控制面组件中有一个kube-controller-manager,这个组件就是Kubernetes中的"资源大管家",也就是一系列的"控制器集合"。我们以一个Deployment的创建过程为例,分析一下其中涉及的主要控制器及其工作过程,如图2-2所示。

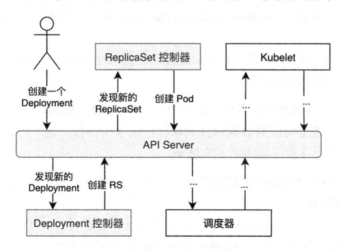

图 2-2　Deployment 的创建过程

在编辑好一个Deployment的YAML配置文件，执行kubectl apply -f ×××-deployment.yaml命令之后，这个资源就被提交到了kube-apiserver。接着kube-controller-manager中的Deployment控制器会收到消息，然后根据Deployment资源的spec定义创建相应的ReplicaSet资源，同样提交给kube-apiserver。紧接着kube-controller-manager中的ReplicaSet控制器又会收到ReplicaSet资源创建的消息，于是根据ReplicaSet资源的spec定义创建相应的Pod资源。最后就是调度器完成Pod的节点绑定，至此Kubelet就完成了对应节点上的Pod创建。当然，后面的过程不是目前我们关注的重点，所以图2-2中也没有详细描述这些过程。

这就是原生控制器的典型工作流程，在Kubernetes中通过各种各样的"资源对象"来描述集群的期望状态，然后通过相应的控制器来完成这个资源对象的调谐逻辑。

2.2 理解 Operator 模式

Operator模式是Kubernetes高度可拓展性的精髓所在，官方文档对Operator模式的介绍可以在https://kubernetes.io/docs/concepts/extend-kubernetes/operator/中找到。

Operator模式让用户能够通过自定义资源来管理自己的应用。这种模式的本质是将一个领域运维人员的运维经验，也就是把他们所维护的应用应该怎么部署、出现异常了应该怎么恢复这一系列运维过程在Kubernetes上的操作程序化，并交给自定义控制器去实施。

我们可以先通过一个例子来了解Operator的工作过程（参考官方示例）：

1）定义一个名为SampleDB的Custom Resource（下文会介绍自定义资源）。

2）通过Deployment方式部署一个Operator的Controller部分。

3）在Controller的代码逻辑中查询apiserver，从而获取SampleDB的具体内容。

4）Operator中的核心逻辑是告诉apiserver怎样让集群的实际状态和自定义资源中声明的状态保持一致，比如：

① 如果新建了一个SampleDB，Operator就会创建一个PersistentVolumeClaim来提供持久化数据库存储，创建一个StatefulSet来运行SampleDB，并创建一个Job来完成初始化配置工作。

② 如果删除了这个SampleDB，Operator相应地打一个快照，然后保证移除前面创建的StatefulSet资源和Volume资源。

5）Operator也可以管理数据库备份，对于每一个SampleDB资源，Operator可以决定在什么时候去运行一个Pod，连上对应的数据库并完成相应的备份操作。这些Pod可以通过ConfigMap或者Secret来获取数据库连接信息和认证信息等。

如果读者是第一次接触Operator相关的概念，可能会对这个例子有些许不解，其中涉及的一些细节我们再多解释几句。

首先,资源是什么？一个资源是Kubernetes API的一个端点,包含一组特定类型对象的集合,比如Pods资源包含Pod对象的集合。在Kubernetes集群中通过kubectl get pods命令可以查询到Pod列表，同样我们常见的Service、Deployment等对象也都对应具体的资源：Services、Deployments

等。资源可以增、删、改、查，对应apiserver代码中定义的某个结构体（实际已经单独拆分出来放到单独的项目里了），如内存中的一个对象、ETCD中的一组数据。所谓自定义资源，就是用户可以提交一个类似Deployment声明的结构定义给Kubernetes，比如上文的SampleDB，所以我们也可以通过kubectl get SampleDB之类的命令来查询SampleDB。而SampleDB中可以定义各种控制具体的数据库应用所需的字段，类似Deployment中的各种属性，比如常见的replicas控制副本数等。

然后Controller能够做的事情就类似于kube-controller-manager，可以获取自定义资源中的声明内容，通过调用apiserver来管理集群的实际状态，比如查询ConfigMap内容来获取数据库连接地址，创建一个StatefulSet来启动数据库实例等。

至此，我们大致了解了Operator能做什么。这里涉及两个核心概念——Custom Resource和Controller，可以通过官方文档进一步了解这些概念：

- Custom Resource：https://kubernetes.io/docs/concepts/extend-kubernetes/api-extension/custom-resources/。
- Controller：https://kubernetes.io/docs/concepts/architecture/controller/。

2.3　Operator 开发环境准备

第1章介绍了如何使用Kind来搭建单节点、3节点Kubernetes环境。本章开始着手Operator开发，在这之前，除了需要准备一套可用的Kubernetes环境之外，读者还需要掌握常用的一些集群运维操作。本书使用Kind工具搭建Kubernetes环境，如果读者对Kind的操作不是很熟悉，可以回过头复习一下1.2.5节。

2.4　Kubebuilder 的安装配置

Kubebuilder是一个用于Operator程序构建和发布的工具。我们先把它安装到本地，然后利用Kubebuilder快速开发一个Hello World性质的Operator程序，这样读者可以快速地对Operator模式有一个直观了解。

如果读者的开发环境是Linux或者amd64的Mac，可以直接通过以下命令安装Kubebuilder。

```
curl -L -o kubebuilder https://go.kubebuilder.io/dl/latest/$(go env GOOS)/$(go env GOARCH)
chmod +x kubebuilder
sudo mv kubebuilder /usr/local/bin/
```

安装完成后，可通过kubebuilder version命令查看一下当前版本的信息，笔者的环境中Kubebuilder是3.2.0版本。

如果读者的开发环境是amd64的Mac，通过上面的方式安装Kubebuilder大概率会失败。我们看一下GitHub release页面（https://github.com/kubernetes-sigs/kubebuilder/releases），如图2-3所示。

图 2-3　Assets 页面

这是2021年10月30日release的3.2.0版本的Assets页面。可以看到这里arm64架构的只有Linux版本，而Darwin系统的又只有amd64版本，所以我们要安装Darwin-Arm64版本的Kubebuilder就只能自己编译安装了，具体的编译和安装代码如下：

```
git clone https://github.com/kubernetes-sigs/kubebuilder.git
cd kubebuilder
git checkout release-3.2.0
make build
make install
```

这里可以选择切换到某个release分支去构建Kubebuilder，当然直接使用main也是可以的，只是有更大的概率会遇到bug。执行完make build命令之后，Kubebuilder可执行文件会存放在当前目录的bin/子目录下，这时可以手动将Kubebuilder复制到/usr/local/bin等由$PATH环境变量指向的目录中，也可以直接执行make install命令将Kubebuilder复制到$GOPATH/bin目录下（前提是需要将$GOPATH/bin加到$PATH中）。

2.5　从 Application Operator Demo 开始

本节将通过一个简单的Demo程序来快速走一遍Operator程序初始化、API定义与实现、打包、本地运行、发布等过程。类比于Deployment可以管理Pod，我们现在尝试开发一个简单的Pod管理控制器Application Operator，但是暂时只支持Pod副本数的管理。

不考虑各种异常处理等逻辑，我们要实现的功能大致如图2-4所示。

我们要通过一个Application类型来定义一个自己的资源对象，然后在控制器中获取这个资源对象的详细配置，接着根据它的配置去创建相应数量的Pod，就像Deployment那样工作。当然，为了避免Demo项目过于复杂，这里不去考虑过多的异常处理、程序健壮性、业务逻辑完备等。

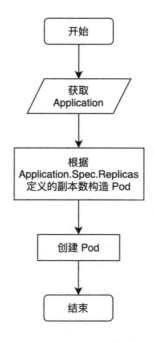

图 2-4　Demo 项目流程

2.5.1　创建项目

在本地家目录下新建一个**MyOperatorProjects**目录用来存放所有的Operator相关项目，命令如下。后续所有演示项目都会存放在这个目录下。

```
cd ~
mkdir MyOperatorProjects/
cd MyOperatorProjects/
```

接着在**MyOperatorProjects**目录下创建我们的第一个项目：

```
# cd ~/MyOperatorProjects/
# mkdir application-operator/
# cd application-operator/
# kubebuilder init --domain=danielhu.cn \
--repo=github.com/daniel-hutao/application-operator \
--owner Daniel.Hu
Writing kustomize manifests for you to edit...
Writing scaffold for you to edit...
Get controller runtime:
$ go get sigs.k8s.io/controller-runtime@v0.10.0
Update dependencies:
$ go mod tidy
Next: define a resource with:
$ kubebuilder create api
```

执行完kubebuilder init ×××命令之后，可以看到以上日志输出。在执行下一步创建API之前，我们先看一下init命令执行完之后Kubebuilder为我们准备了什么：

```
application-operator/
|-- config
|   |-- default
|   |   |-- kustomization.yaml
|   |   |-- manager_auth_proxy_patch.yaml
|   |   `-- manager_config_patch.yaml
|   |-- manager
|   |   |-- controller_manager_config.yaml
|   |   |-- kustomization.yaml
|   |   `-- manager.yaml
|   |-- prometheus
|   |   |-- kustomization.yaml
|   |   `-- monitor.yaml
|   `-- rbac
|       |-- auth_proxy_client_clusterrole.yaml
|       |-- auth_proxy_role_binding.yaml
|       |-- auth_proxy_role.yaml
|       |-- auth_proxy_service.yaml
|       |-- kustomization.yaml
|       |-- leader_election_role_binding.yaml
|       |-- leader_election_role.yaml
|       |-- role_binding.yaml
|       `-- service_account.yaml
|-- hack
|   `-- boilerplate.go.txt
|-- Dockerfile
|-- go.mod
|-- go.sum
|-- main.go
|-- Makefile
`-- PROJECT
```

这里我们暂时不要陷入细节，避免迷失在复杂的内容中导致失去耐心。先简单看一下几个主要文件中都有什么内容，看看刚才传给kubebuilder命令的几个参数最后都起了哪些作用，从而在心中建立一个初步的感性认知。

- PROJECT：这个文件中可以看到项目的一些元数据，比如domain、projectName、repo等信息。回头添加了API之后，可以留意一下这个文件会发生什么变化。

```
domain: danielhu.cn
layout:
- go.kubebuilder.io/v3
projectName: application-operator
repo: github.com/daniel-hutao/application-operator
version: "3"
```

- main.go：这个文件内容不少，大家可以简单浏览一遍这些代码，但是不要纠结每一行的含义，其中有很多知识点还没有学习到。不过不要担心，后续章节会逐步介绍这里的所有细节。现在我们不着急讲解main的代码逻辑，先看这里的一个有趣的特性：每个文件头Copyright会加上--owner参数的内容。

```
/*
Copyright 2022 Daniel.Hu.
...
```

这个Copyright其实来自hack/boilerplate.go.txt文件。

- config：这个目录中放置了很多个YAML文件。其中包括RBAC权限相关的YAML文件、Prometheus监控服务发现（ServiceMonitor）相关的YAML文件、控制器（Manager）本身部署的YAML文件等。
- Dockerfile：最终Operator程序编译、构建镜像的逻辑就在这里，我们可以通过修改这个文件来解决一些镜像构建相关的问题，比如Go语言依赖下载的默认GOPROXY配置在国内不一定能访问到，可以在其中配置国内的proxy地址等。
- Makefile：这是解放劳动力的工具，这里实现了通过make ×××轻松实现整个程序的编译构建、镜像推送、部署、卸载等操作，后面会经常用到。

为了方便跟踪接下来每个命令的执行结果，可以利用git来跟踪代码变更：

```
# git init
Initialized empty Git repository in /Users/danielhu/MyOperatorProjects/application-operator/.git/
# git add .
# git commit -m "Init project"
[master (root-commit) df6c019] Init project
 26 files changed, 1451 insertions(+)
...
```

接着执行的每一步操作都可以利用git来跟踪发生了哪些变更，也可以利用git来轻松还原不小心改错的配置或代码。

2.5.2 添加 API

在application-operator目录下继续执行命令，连续输入两次y：

```
# kubebuilder create api \
--group apps --version v1 --kind Application
Create Resource [y/n]
y
Create Controller [y/n]
y
Writing kustomize manifests for you to edit...
Writing scaffold for you to edit...
api/v1/application_types.go
controllers/application_controller.go
Update dependencies:
$ go mod tidy
Running make:
$ make generate
go: creating new go.mod: module tmp
...
```

```
go get: added sigs.k8s.io/controller-tools v0.7.0
go get: added sigs.k8s.io/structured-merge-diff/v4 v4.1.2
go get: added sigs.k8s.io/yaml v1.2.0
/Users/danielhu/MyOperatorProjects/application-operator/bin/controller-gen
object:headerFile="hack/boilerplate.go.txt" paths="./..."
Next: implement your new API and generate the manifests (e.g. CRDs,CRs) with:
$ make manifests
```

这时Kubebuilder又为我们做了什么呢？执行git status命令：

```
# git status
On branch master
Changes not staged for commit:
  (use "git add <file>..." to update what will be committed)
  (use "git checkout -- <file>..." to discard changes in working directory)

        modified:   PROJECT
        modified:   go.mod
        modified:   main.go

Untracked files:
  (use "git add <file>..." to include in what will be committed)

        api/
        config/crd/
        config/rbac/application_editor_role.yaml
        config/rbac/application_viewer_role.yaml
        config/samples/
        controllers/

no changes added to commit (use "git add" and/or "git commit -a")
```

可以看到新增的文件是：

- api/。
- config/crd/。
- config/rbac/application_editor_role.yaml。
- config/rbac/application_viewer_role.yaml。
- config/samples/。
- controllers/。

变更的文件是：

- PROJECT。
- go.mod。
- main.go。

我们从变更的文件开始看一下create api的效果。

- PROJECT：多了resources部分的信息，如以下加粗部分所示：

```
domain: danielhu.cn
layout:
```

```
- go.kubebuilder.io/v3
projectName: application-operator
repo: github.com/daniel-hutao/application-operator
resources:
- api:
    crdVersion: v1
    namespaced: true
  controller: true
  domain: danielhu.cn
  group: apps
  kind: Application
  path: github.com/daniel-hutao/application-operator/api/v1
  version: v1
version: "3"
```

- main.go：查看其中多的几行代码，先留个印象，暂时不细究：

```
import(
    //...
    appsv1 "github.com/daniel-hutao/application-operator/api/v1"
    "github.com/daniel-hutao/application-operator/controllers"
)
// ...
func main() {
    // ...
    mgr, err := ctrl.NewManager(ctrl.GetConfigOrDie(), ctrl.Options{
        Scheme:                 scheme,
        MetricsBindAddress:     metricsAddr,
        Port:                   9443,
        HealthProbeBindAddress: probeAddr,
        LeaderElection:         enableLeaderElection,
        LeaderElectionID:       "ad4dd4db.danielhu.cn",
    })
    if err != nil {
        setupLog.Error(err, "unable to start manager")
        os.Exit(1)
    }

    if err = (&controllers.ApplicationReconciler{
        Client: mgr.GetClient(),
        Scheme: mgr.GetScheme(),
    }).SetupWithManager(mgr); err != nil {
        setupLog.Error(err, "unable to create controller", "controller", "Application")
        os.Exit(1)
    }
    // ...
    if err := mgr.Start(ctrl.SetupSignalHandler()); err != nil {
        setupLog.Error(err, "problem running manager")
        os.Exit(1)
    }
}
```

- go.mod：其中多出来的是BDD测试相关的ginkgo和gomega依赖。

接下来看一下新增的几个文件/目录：

- api/：这个目录中包含刚才添加的API，后面会经常编辑这里的application_types.go文件。
- config/crd：存放的是crd部署相关的kustomize文件。后面会有专门的章节介绍kustomize的用法。
- config/rbac：其中多了两个文件：
    - application_editor_role.yaml：定义了一个有applications资源编辑权限的ClusterRole。
    - application_viewer_role.yaml：定义了一个有applications资源查询权限的ClusterRole。
- samples/apps_v1_application.yaml：这是一个CR示例文件，从这个文件的骨架很容易看到通过填充内容即可用来创建一个自定义资源Application类型的实例：

```yaml
apiVersion: apps.danielhu.cn/v1
kind: Application
metadata:
  name: application-sample
spec:
  # TODO(user): Add fields here
```

- controllers/：这里包含控制器的代码逻辑入口。我们看一下Reconcile函数，"调谐"（Reconcile）这个词会贯穿整本书。

```go
func (r *ApplicationReconciler) Reconcile(ctx context.Context, req ctrl.Request) (ctrl.Result, error) {
    _ = log.FromContext(ctx)
    // TODO(user): your logic here
    return ctrl.Result{}, nil
}
```

2.5.3　CRD 实现

我们前面提到过CRD的代码主要定义在api/v1/application_types.go文件中，先打开看一下现有代码是怎样的。主要看Spec结构，也就是ApplicationSpec这个结构体的定义：

```go
type ApplicationSpec struct {
    Foo string `json:"foo,omitempty"`
}
```

这里包含一个示例字段Foo，需要删除这个字段，添加自己需要的配置：

```go
import (
    corev1 "k8s.io/api/core/v1"
    metav1 "k8s.io/apimachinery/pkg/apis/meta/v1"
)
...
type ApplicationSpec struct {
    Replicas int32                 `json:"replicas,omitempty"`
    Template corev1.PodTemplateSpec `json:"template,omitempty"`
}
```

这里使用两个属性：

- Replicas：用于声明Pod副本的数量。
- Template：用于声明Pod模板的配置。

2.5.4 CRD 部署

修改好application_types.go中的ApplicationSpec之后，通过make manifests命令来生成ClusterRole和CustomResourceDefinition配置。

```
# make manifests
/Users/danielhu/MyOperatorProjects/application-operator/bin/controller-gen
rbac:roleName=manager-role crd webhook paths="./..."
output:crd:artifacts:config=config/crd/bases
```

同样可以通过git来查看这个命令执行后的效果：

```
# git status
On branch master
Untracked files:
  (use "git add <file>..." to include in what will be committed)

        config/crd/bases/
        config/rbac/role.yaml

nothing added to commit but untracked files present (use "git add" to track)
```

看一下这两个文件/目录：

- config/crd/bases/：这个目录中新增了一个apps.danielhu.cn_applications.yaml文件，也就是Application类型的CRD配置：

```
---
apiVersion: apiextensions.k8s.io/v1
kind: CustomResourceDefinition
metadata:
  annotations:
    controller-gen.kubebuilder.io/version: v0.7.0
  creationTimestamp: null
  name: applications.apps.danielhu.cn
spec:
  group: apps.danielhu.cn
  names:
    kind: Application
    listKind: ApplicationList
    plural: applications
    singular: application
  scope: Namespaced
  versions:
  - name: v1
    ...
```

这里有很多熟悉的字段，如以上加粗部分所示，我们可以很容易辨认出这个配置文件的含义。

- config/rbac/role.yaml：这里面定义的是一个ClusterRole，从名字manager-role上大致可以猜到，这是后面Controller部署后将充当的"角色"。该文件定义了对applications资源的创建、删除、查询、更新等操作。

最后，我们可以通过make install命令来完成CRD的部署过程：

```
# make install
/Users/danielhu/MyOperatorProjects/application-operator/bin/controller-gen
rbac:roleName=manager-role crd webhook paths="./..."
output:crd:artifacts:config=config/crd/bases
    /Users/danielhu/MyOperatorProjects/application-operator/bin/kustomize build
config/crd | kubectl apply -f -
    customresourcedefinition.apiextensions.k8s.io/applications.apps.danielhu.cn
created
```

当然，make install的执行是包含make manifests的，不过我们还是分开操作以方便讲解其中具体的过程。执行完make install之后，CRD也就应用到我们的Kubernetes集群里了。这时可以通过kubectl命令来查看刚才创建的CRD：

```
# kubectl get crd
NAME                              CREATED AT
applications.apps.danielhu.cn     2022-01-01T09:28:30Z
# kubectl get application
No resources found in default namespace.
```

可以看到，applications.apps.danielhu.cn这个CRD已经存在了。这时就像查询Deployment时可以用kubectl get deployment一样，可以通过kubectl get application来查询刚才自定义的资源。这里的报错信息是没找到资源，而不是"error: the server doesn't have a resource type "application""。换言之，此时kube-apiserver已经能识别这种资源了，所以接下来可以创建一个具体的Application类型的资源对象实例。

2.5.5 CR 部署

和写一个Deployment资源配置对应的×××.yaml一样，创建Application同样需要一个YAML文件。我们在前面提到过在config/samples下有一个apps_v1_application.yaml，其中已经有一些默认配置项了，我们稍微修改一下，加上自定义的属性：

```yaml
apiVersion: apps.danielhu.cn/v1
kind: Application
metadata:
  name: application-sample
  namespace: default
  labels:
    app: nginx
spec:
```

```yaml
  replicas: 3
  template:
    spec:
      containers:
        - name: nginx
          image: nginx:1.14.2
          ports:
            - containerPort: 80
```

然后执行apply再看一下结果如何：

```
# kubectl apply -f config/samples/apps_v1_application.yaml
application.apps.danielhu.cn/application-sample created
```

这时可以在集群中查到刚才创建的Application：

```
# kubectl get application
NAME                  AGE
application-sample    1m19s
```

很和谐的输出，一个Application类型的application-sample实例创建出来了，其中包含自定义的所有字段。但是这时并不会有Pod被自动创建出来，因为还没有实现相应的控制器逻辑。

2.5.6　Controller 实现

现在已经有了CR，该实现Controller逻辑去创建Pod了。在controllers/ application_controller.go中的Reconcile()方法内添加如下代码：

```go
func (r *ApplicationReconciler) Reconcile(ctx context.Context, req ctrl.Request)
(ctrl.Result, error) {
    l := log.FromContext(ctx)

    // get the Application
    app := &dappsv1.Application{}
    if err := r.Get(ctx, req.NamespacedName, app); err != nil {
        if errors.IsNotFound(err) {
            l.Info("the Application is not found")
            return ctrl.Result{}, nil
        }
        l.Error(err, "failed to get the Application")
        return ctrl.Result{RequeueAfter: 1 * time.Minute}, err
    }

    // create pods
    for i := 0; i < int(app.Spec.Replicas); i++ {
        pod := &corev1.Pod{
            ObjectMeta: metav1.ObjectMeta{
                Name:      fmt.Sprintf("%s-%d", app.Name, i),
                Namespace: app.Namespace,
                Labels:    app.Labels,
            },
            Spec: app.Spec.Template.Spec,
        }
```

```go
        if err := r.Create(ctx, pod); err != nil {
            l.Error(err, "failed to create Pod")
            return ctrl.Result{RequeueAfter: 1 * time.Minute}, err
        }
        l.Info(fmt.Sprintf("the Pod (%s) has created", pod.Name))
    }

    l.Info("all pods has created")
    return ctrl.Result{}, nil
}
```

这时import部分大致如下：

```go
import (
    "context"
    "fmt"
    "time"

    corev1 "k8s.io/api/core/v1"
    "k8s.io/apimachinery/pkg/api/errors"
    metav1 "k8s.io/apimachinery/pkg/apis/meta/v1"
    "k8s.io/apimachinery/pkg/runtime"
    ctrl "sigs.k8s.io/controller-runtime"
    "sigs.k8s.io/controller-runtime/pkg/client"
    "sigs.k8s.io/controller-runtime/pkg/log"

    dappsv1 "github.com/daniel-hutao/application-operator/api/v1"
)
```

这里我们把自己的api/v1打包并使用别名dappsv1，避免和k8s.io/api/apps/v1包产生混淆。我们分析一下上面的代码：

```go
// 声明一个*Application类型的实例app用来接收我们的CR
app := &dappsv1.Application{}
// NamespacedName在这里也就是default/application-sample
if err := r.Get(ctx, req.NamespacedName, app); err != nil {
    // err分很多种情况，如果找不到，一般不需要进一步处理，只是说明这个CR被删了而已
    if errors.IsNotFound(err) {
        l.Info("the Application is not found")
        // 直接返回，不带错误，结束本次调谐
        return ctrl.Result{}, nil
    }
    // 除了NotFound之外的错误，比如连不上apiserver等，这时需要打印错误信息，然后返回这个错误以及表示1分钟后重试的Result
    l.Error(err, "failed to get the Application")
    return ctrl.Result{RequeueAfter: 1 * time.Minute}, err
}
```

接下来就可以启动这个Controller来看一下效果了。

2.5.7 启动 Controller

直接执行make run即可以运行代码，在终端可以看到如下日志输出（删除了业务无关的日志）：

```
2022-01-02T14:12:02.696+0800    INFO    controller-runtime.metrics    metrics
server is starting to listen    {"addr": ":8080"}
2022-01-02T14:12:02.696+0800    INFO    setup    starting manager
2022-01-02T14:12:02.696+0800    INFO    controller.application    Starting
EventSource    {"reconciler group": "apps.danielhu.cn", "reconciler kind":
"Application", "source": "kind source: /, Kind="}
2022-01-02T14:12:02.696+0800    INFO    starting metrics server {"path":
"/metrics"}
2022-01-02T14:12:02.696+0800    INFO    controller.application    Starting
Controller    {"reconciler group": "apps.danielhu.cn", "reconciler kind":
"Application"}
2022-01-02T14:12:02.798+0800    INFO    controller.application    Starting workers
{"reconciler group": "apps.danielhu.cn", "reconciler kind": "Application", "worker
count": 1}
2022-01-02T14:12:02.806+0800    INFO    controller.application    the Pod
(application-sample-0) has created    {"reconciler group": "apps.danielhu.cn",
"reconciler kind": "Application", "name": "application-sample", "namespace":
"default"}
2022-01-02T14:12:02.820+0800    INFO    controller.application    the Pod
(application-sample-1) has created    {"reconciler group": "apps.danielhu.cn",
"reconciler kind": "Application", "name": "application-sample", "namespace":
"default"}
2022-01-02T14:12:02.832+0800    INFO    controller.application    the Pod
(application-sample-2) has created    {"reconciler group": "apps.danielhu.cn",
"reconciler kind": "Application", "name": "application-sample", "namespace":
"default"}
2022-01-02T14:12:02.832+0800    INFO    controller.application    all pods has
created    {"reconciler group": "apps.danielhu.cn", "reconciler kind": "Application",
"name": "application-sample", "namespace": "default"}
```

每一行有时间信息和结构化日志信息,我们略过这些信息后,看一下纯日志文本中有哪些信息:

```
metrics server is starting to listen
starting manager
Starting EventSource
starting metrics server {"path": "/metrics"}
Starting Controller
Starting workers
the Pod (application-sample-0) has created
the Pod (application-sample-1) has created
the Pod (application-sample-2) has created
all pods has created
```

前面是"框架"抛出的一些日志,我们还是可以从中看到很多信息,包括控制器的大致启动逻辑等。加粗部分内容就是我们自己在Reconcile函数中自定义的日志内容。从日志看我们的代码没有任何异常,所以继续看一下Pod是不是真的已经创建出来了:

```
# kubectl get pod
NAME                      READY   STATUS    RESTARTS   AGE
application-sample-0      1/1     Running   0          1m4s
application-sample-1      1/1     Running   0          1m4s
```

```
application-sample-2   1/1      Running   0          1m4s
```

2.5.8 部署 Controller

前面我们部署了一个Application类型的资源对象实例,也运行Controller看到了相应副本数的Pod被创建出来。现在我们进一步模拟Operator实际使用时的部署方式,把Controller打包后用容器化方式部署到Kubernetes集群环境中:

```
# 构建镜像
make docker-build IMG=application-operator:v0.0.1
# 推到kind环境
kind load docker-image application-operator:v0.0.1 --name dev
# 部署控制器
make deploy IMG=application-operator:v0.0.1
```

这里可能会遇到几个问题,第一个是依赖下载失败的问题,因为本地GOPROXY配置在容器内是不管用的,这里的编译过程在golang:1.16镜像启动的容器中进行,我们在Dockerfile中可以看到相关逻辑。所以如果遇到与连不上proxy.golang.org相关的错误,比如下面这样:

```
go: github.com/onsi/ginkgo@v1.16.4: Get
"https://proxy.golang.org/github.com/onsi/ginkgo/@v/v1.16.4.mod": dial tcp
142.251.43.17:443: connect: connection refused
```

可以通过在Dockerfile中添加如下命令(加粗行)的方式来解决(goproxy.io和goproxy.cn都是不错的代理,多数场景下两者都能较好地工作。如果遇到有一个连不上,则可以切换到另一个再测试一下):

```
# Build the manager binary
FROM golang:1.16 as builder
ENV GOPROXY=https://goproxy.io
WORKDIR /workspace
```

第二个问题是base镜像下载失败的问题,看Dockerfile可以发现在22行左右有这样一行命令:

```
FROM gcr.io/distroless/static:nonroot
```

很明显,这个镜像在国内网络环境下默认是下载不到的,我们可以折中一下,在本地执行如下命令从其他库中转一份:

```
docker pull kubeimages/distroless-static
docker tag kubeimages/distroless-static:latest \
    gcr.io/distroless/static:nonroot
```

第三个问题是可能遇到访问raw.githubusercontent.com有问题,错误信息如下:

```
test -f /Users/danielhu/go/code/application-operator/testbin/setup-envtest.sh ||
curl -sSLo /Users/danielhu/go/code/application-operator/testbin/setup-envtest.sh
https://raw.githubusercontent.com/kubernetes-sigs/controller-runtime/v0.8.3/hack/se
tup-envtest.sh
  curl: (35) LibreSSL SSL_connect: SSL_ERROR_SYSCALL in connection to
raw.githubusercontent.com:443
```

这时可以把curl命令复制出来试试，或者直接上GitHub，到kubernetes-sigs/controller-runtime项目中找到setup-envtest.sh，下载到本地并放在项目的testbin/下。如果访问GitHub都不行，那就是网络环境太苛刻了。不过在新版的Kubebuilder中已经不需要这个脚本了，而是直接使用Go语言实现这个功能，所以如果大家下载的Kubebuilder版本是最新的，应该不会遇到这个问题。

第四个问题是部署之后可以看到这个Pod中有一个容器所用的镜像gcr.io/kubebuilder/kube-rbac-proxy:v0.8.0，很明显这个镜像需要中转一下（不要忘记这个镜像是在Kind环境中使用的，最后需要加载一下）：

```
docker pull kubesphere/kube-rbac-proxy:v0.8.0
docker tag kubesphere/kube-rbac-proxy:v0.8.0 \
gcr.io/kubebuilder/kube-rbac-proxy:v0.8.0
kind load docker-image gcr.io/kubebuilder/kube-rbac-proxy:v0.8.0 --name dev
```

如果是M1的MacBook，还可能会在执行make test/make docker-build的时候遇到这个问题：

```
/Users/danielhu/MyOperatorProjects/application-operator/bin/controller-gen
rbac:roleName=manager-role crd webhook paths="./..."
output:crd:artifacts:config=config/crd/bases
/Users/danielhu/MyOperatorProjects/application-operator/bin/controller-gen
object:headerFile="hack/boilerplate.go.txt" paths="./..."
    go fmt ./...
    go vet ./...
unable to find a version that was supported for platform darwin/arm64
```

arm64迟早会被支持，但是目前来看还是有不少兼容性的问题。我们可以通过如下方式暂且绕开：

找到Makefile中的这行代码：

```
KUBEBUILDER_ASSETS="$(shell $(ENVTEST) use $(ENVTEST_K8S_VERSION) -p path)" go test ./... -coverprofile cover.out
```

然后改成这样：

```
KUBEBUILDER_ASSETS="$(shell $(ENVTEST) --arch=amd64 use $(ENVTEST_K8S_VERSION) -p path)" go test ./... -coverprofile cover.out
```

也就是加上--arch=amd64这个参数。

大概率会遇到的问题基本上就是这几个，大家实际操作的环境可能会有一定的差异，使用的Kubebuilder等工具的版本也会有一定的区别，所以可能还会遇到其他错误，不过基本都可以通过"仔细看日志"和"Google一下"来解决。

顺利执行完前面的几个命令，就可以在环境中看到Operator程序了：

```
# kubectl get pod -n application-operator-system
NAME                                                    READY
STATUS     RESTARTS    AGE
application-operator-controller-manager-7d57898455-j6rp2 2/2    Running    0    2m
```

至此，就实现了一个非常简单的Operator程序的开发、编译构建、打包、部署等流程。当然，如果前面执行make run启动了控制器创建出来的3个Pod没有被删除，这时再次部署一个Operator控制器，日志里肯定会有报错信息的，因为我们没有实现Pod存在校验等代码逻辑。不过这不是重点，有没有这些错误，我们的Demo程序都算走完了。到这里，大家基本体验了一遍Operator的快速开发流程。

2.5.9 资源清理

前面我们通过几个步骤分别部署了CRD、CR和Controller，这时环境中新增的资源不少（各种rbac相关的、部署相关的资源），如果这些资源都用kubectl一个一个来清理，还是很痛苦的。同样，清理的步骤也可以通过Makefile来实现：

```
# 卸载Controller
make undeploy
# 卸载CRD
make uninstall
```

当前我们使用的Kubebuilder版本在执行make undeploy的时候还会遇到一个错误：

```
Error: invalid argument "no" for "--ignore-not-found" flag: strconv.ParseBool: parsing "no": invalid syntax
```

这是一个挺明显的bug，我们在Makefile中大致找一下关键字就可以很快定位到这段有问题的代码（第82行左右）：

```
ifndef ignore-not-found
  ignore-not-found = no
endif
```

改成这样：

```
ifndef ignore-not-found
  ignore-not-found = false
endif
```

这个错误在本书出版时应该已经被修复了，但是也保不准又出现了其他新的bug。我们研究开源软件的过程就是这样，不断遇到问题，不断解决问题，这个过程也许会很痛苦，但是问题解决的那一刻会很有成就感。只要不放弃，在被各种"问题"洗礼的过程中，自己的能力会越来越强。

2.6 Operator的发展历史

前面我们通过一个Demo程序快速过了一遍Operator项目的编码、打包、部署等流程，本节我们来了解Operator概念的由来与发展历程等背景知识。

2.6.1 Operator概念的提出

Operator概念由CoreOS的CTO Brandon Philips在2016年提出，原文是 *Introducing Operators*：

Putting Operational Knowledge into Software，最初链接是 https://coreos.com/blog/introducing-operators.html，不过现在这个站点已经重定向到了 https://cloud.redhat.com/blog，因而这篇文章也不好找了。其中有这么几句话，意译过来的意思是：

SRE通过编写运维软件来运维应用，他们是工程师，也是开发者，知道怎么针对特定应用领域来开发运维软件，这些运维软件中包含特定应用领域的运维经验。

我们把这种新的软件类型叫作Operator。一个Operator是特定应用的控制器，通过拓展Kubernetes API来创建、配置和管理复杂有状态的应用实例，代替用户人工操作。它构建在Kubernetes Resource和Controller概念之上，同时包含领域或应用特定的知识，从而自动化地实现通用的运维任务。

2.6.2 第一个 Operator 程序

时至今日，Operator方式部署已经成为分布式应用的交付标准，几乎所有的主流分布式应用都有其对应的Kubernetes Operator部署项目。第一个Operator程序是2016年CoreOS推出的etcd-operator，代码库地址是https://github.com/coreos/etcd-operator。

我们知道Kubernetes作为一个容器编排系统，其核心机制是"控制器模式"，当API资源发生变化时，控制器开始各种逻辑处理，完成所需的编排操作。所以我们能不能通过自己实现控制器来管理Kubernetes中的资源呢？2016年秋天的一个原本普通的周四，来自CoreOS的邓洪超相约去同事家里一起结对编程。这两位工程师最早思考这种开发模式并且将其实现，这个项目就是后来广为人知的etcd-operator。

我们要实现在Kubernetes中用Operator方式管理一个分布式应用，第一步是定义一个"自定义对象"来描述这个分布式应用，比如通过一个叫作EtcdCluster的对象来描述ETCD集群的期望状态，然后通过控制器来将集群中ETCD的状态"调谐"到期望的状态。那么怎么描述这种状态呢？Kubernetes中原生支持的对象是Pod、Deployment，显然没有一个对象叫作EtcdCluster，不过当时Kubernetes项目已经有了允许用户添加自定义API对象的能力，也就是CRD的前身TPR（Third Party Resource，第三方资源）。

有了TPR可以用来描述EtcdCluster之后，剩下的事情就很清晰了，即在自定义控制器中实现EtcdCluster对象的增删改查，实现ETCD实例的相应编排逻辑。到此，第一版etcd-operator完成，世界上第一个Kubernetes Operator项目就此诞生！

2.6.3 Operator 的崛起

在etcd-operator诞生几个月后，邓洪超和同事一起在KubeCon中展示了一次Demo，推荐大家尝试使用Operator模式来部署ETCD集群。当时这个Demo在社区引起了很大的反响，很多开发者开始尝试用这种全新的模式来发布自己的分布式应用。随着Prometheus等原先很难在Kubernetes上运行的应用纷纷宣布实现Operator化部署之后，大家已经逐步意识到Operator模式将会成为有状态应用上Kubernetes的标准。

但是"控制器模式"在Google团队看来应该是隐藏在Kubernetes内部的一种机制，不应该直接开放给用户。Operator模式将控制器模式从kube-controller-manager中剥离出来，直接暴露给了开发者。另外，Kubernetes项目的发起人之一Brendan Burns，也是TPR特性的最初发起人之一，那时突然宣布离开Google加入微软，这件事进一步激化了Google团队和Operator之间的矛盾，于是在2017年年初，Google开始和Red Hat一起在社区推出UAS（User Aggregated APIServer），也就是后来的APIServer Aggregator。Google和Red Hat在社区呼吁通过UAS取代TPR，一度让大家以为Operator这个新生事物要被扼杀在襁褓里了。

这时Brandon Philips，也就是第一个提出Operator概念的CoreOS的CTO，在GitHub上开了一个Gist，让社区里所有的开发者发声，留下自己正在使用Operator的项目和链接，希望能够一起挽救Operator。大家可以进Gist感受一下这段历史：https://gist.github.com/philips/a97a143546c87b86b870a82a753db14c。现在进去的话可以看到项目列表已经被移到了https://github.com/coreos/awesome-kubernetes-extensions维护，然后awesome-kubernetes-extensions又被移到了https://github.com/operator-framework/awesome-operators，当然目前awesome-operators也已经成为历史，Operator项目已经太多了，目前大家已经直接在https://operatorhub.io上共享Operator。伴随着巨大的社区压力，面对众多开发者建立起来的Operator生态，Google和Red Hat最终选择了妥协与让步。几个月后，Google和Red Hat的工程师在社区宣布使用CRD替换TPR，其实两者的区别并不大。至此，Operator终于度过了危险期，存活了下来，并在此后几年里不断发展壮大。

2.7 本章小结

本章我们从什么是Operator模式开始介绍，然后介绍Kubebuilder工具的安装，接着通过一个简单的Application Operator Demo来介绍Operator开发的整体过程，最后介绍了Operator的发展历史，希望能够让大家对Operator是什么、能干什么、怎么来的等有一个宏观理解。

Kubernetes Operator 开发进阶

第二篇

进 阶

第 3 章
Kubernetes API 介绍

我们开发的 Operator 应用本质上还是在和 Kubernetes API 打交道，所以有必要进一步学习 Kubernetes API 的相关知识。

3.1 认识 Kubernetes API

我们知道 Kubernetes 集群中有一个 kube-apiserver 组件，所有组件需要操作集群资源时都通过调用 kube-apiserver 提供的 RESTful 接口来实现。kube-apiserver 进一步和 ETCD 交互，完成资源信息的更新，另外 kube-apiserver 也是集群内唯一和 ETCD 直接交互的组件。

Kubernetes 中的资源本质就是一个 API 对象，这个对象的"期望状态"被 API Server 保存在 ETCD 中，然后提供 RESTful 接口用于更新这些对象。我们可以直接和 API Server 交互，使用"声明"的方式来管理这些资源（API 对象），也可以通过 kubectl 这种命令行工具，或者 client-go 这类 SDK。当然，在 Dashboard 上操作资源的方式和 API Server 交互也是一种方式，不过这不是我们关注的重点。

我们可以在 Kubernetes 的官网中找到 API 文档：https://kubernetes.io/docs/reference/generated/kubernetes-api/v1.23/。

3.2 使用 Kubernetes API

我们以一个 Deployment 资源的增删改查操作为例，看一下如何与 API Server 交互。

3.2.1 Curl 方式访问 API

1. 准备工作

由于 kube-apiserver 默认提供的是 HTTPS 服务，而且是双向 TLS 认证，而我们目前的关注重点是API本身，因此先通过kubectl来代理API Server服务：

```
# kubectl proxy --port=8080
Starting to serve on 127.0.0.1:8080
```

这时就可以通过简单的HTTP请求来和API Server交互了：

```
# curl localhost:8080/version
{
  "major": "1",
  "minor": "22",
  "gitVersion": "v1.22.0",
  "gitCommit": "c2b5237ccd9c0f1d600d3072634ca66cefdf272f",
  "gitTreeState": "clean",
  "buildDate": "2021-08-04T20:04:38Z",
  "goVersion": "go1.16.6",
  "compiler": "gc",
  "platform": "linux/arm64"
}%
```

我们还需要一个配置文件来描述Deployment资源，在本地创建一个nginx-deploy.yaml文件，内容如下：

```
apiVersion: apps/v1
kind: Deployment
metadata:
  name: nginx-deploy
spec:
  replicas: 3
  selector:
    matchLabels:
      app: nginx
  template:
    metadata:
      labels:
        app: nginx
    spec:
      containers:
      - name: nginx
        image: nginx:1.14
        ports:
        - containerPort: 80
```

接着就可以开始通过API调用来操作这个Deployment资源了。

2. 资源创建

Deployment的创建API是：

```
POST /apis/apps/v1/namespaces/{namespace}/deployments
```

所以执行下面的命令在default命名空间下创建一个Deployment：

```
# curl -X POST \
-H 'Content-Type: application/yaml' \
--data-binary '@nginx-deploy.yaml' \
http://localhost:8080/apis/apps/v1/namespaces/default/deployments
{
  "kind": "Deployment",
  "apiVersion": "apps/v1",
  "metadata": {
    "name": "nginx-deploy",
    "namespace": "default",
    "uid": "69d6bbde-b7aa-4e79-99ac-b29c2add3fcf",
    "resourceVersion": "275105",
    "generation": 1,
"creationTimestamp": "2022-01-07T12:31:55Z",
...
```

可以看到API Server会响应一个新创建的资源的详细对象描述（采用JSON格式）。这时我们可以通过kubectl来查询这个资源对象：

```
# kubectl get deployment
NAME            READY   UP-TO-DATE   AVAILABLE   AGE
nginx-deploy    3/3     3            3           3m
```

3. 资源删除

和Deployment的创建类似，同样可以利用Kubernetes API来删除资源。Deployment的删除API是：

```
DELETE /apis/apps/v1/namespaces/{namespace}/deployments/{name}
```

所以我们可以执行下面的命令删除前面在default命名空间下创建的Deployment：

```
# curl -X DELETE -H 'Content-Type: application/yaml' \
--data '
gracePeriodSeconds: 0
orphanDependents: false
' \
http://localhost:8080/apis/apps/v1/namespaces/default/deployments/nginx-deploy
{
  "kind": "Status",
  "apiVersion": "v1",
  "metadata": {
  },
  "status": "Success",
  "details": {
    "name": "nginx-deploy",
    "group": "apps",
    "kind": "deployments",
    "uid": "def3da00-e322-4b61-91b4-56125f7eaa83"
  }
}%
```

删除操作成功的响应体是一个Status类型的对象，我们可以很清晰地看到这次被删除的对象所对应的group、kind、name、uid等详细信息。

3.2.2　kubectl raw 方式访问 API

前面我们通过kubectl proxy命令来暴露API Server服务，然后通过curl命令去创建和删除Deployment类型的资源实例。本小节介绍另一种方式，也就是直接使用kubectl xxx --raw来访问API。

```
# kubectl get --raw /version
{
  "major": "1",
  "minor": "22",
  "gitVersion": "v1.22.0",
  "gitCommit": "c2b5237ccd9c0f1d600d3072634ca66cefdf272f",
  "gitTreeState": "clean",
  "buildDate": "2021-08-04T20:04:38Z",
  "goVersion": "go1.16.6",
  "compiler": "gc",
  "platform": "linux/arm64"
}%
```

如以上代码所示，通过kubectl get --raw可以实现和curl类似的效果，只是我们不需要指定API Server的地址了，同样认证信息也不需要，这时默认用了kubeconfig中的连接信息。如果前面创建的Deployment未被删除，在可以通过这种方式来查询nginx- deployment：

```
# kubectl get --raw /apis/apps/v1/namespaces/default/deployments/nginx-deploy
{
  "kind":"Deployment",
  "apiVersion":"apps/v1",
  "metadata":{
    "name":"nginx-deploy",
    "namespace":"default",
    ...
    },
   ...
  "spec":{
    "replicas":3,
    ...
  },
  "status":{
    ...
  }
}
```

这里省略了很多字段，因为我们只关心这个命令的执行结果是不是能够查询到nginx-deploy这个Deployment。kubectl get --raw这种方式和kubectl proxy + curl本质是一样的，只是不需要再开两个窗口分别执行两条命令了。

3.3 理解 GVK：组、版本与类型

其实我们前面已经多次接触到 GVK 这个词了，望文生义，大家不难猜到 GVK 就是 Group、Version、Kind 三个词的首字母缩写。本节详细学习一下 GVK 相关的概念。

我们在描述 Kubernetes API 时经常会用到这样一个四元组：Groups、Versions、Kinds 和 Resources。它们具体是什么含义呢？

- Groups 和 Versions：一个 Kubernetes API Group 表示的是一些相关功能的集合，比如 apps 这个 Group 里面就包含 deployments、replicasets、daemonsets、statefulsets 等资源，这些资源都是应用工作负载相关的，也就放在了同一个 Group 下。一个 Group 可以有一个或多个 Versions，不难理解这里的用意，毕竟随着时间的推移，一个 Group 中的 API 难免有所变化。也许大家已经注意到，以前的 Kubernetes 版本创建 Deployment 时 apiVersion 用过 apps/v1beta1 和 apps/v1beta2，现在已经是 apps/v1 了。

- Kinds 和 Resources：每个 group-version（确定版本的一个组）中都包含一个或多个 API 类型，这些类型就是这里说的 Kinds。每个 Kind 在不同的版本中一般会有所差异，但是每个版本的 Kind 要能够存储其他版本 Kind 的资源类型，无论是通过存储在字段里实现还是通过存储在注解中实现，具体的机制后面会详细讲解。这也就意味着使用老版本的 API 存储新版本类型数据不会引起数据丢失或污染。至于 Resources，指的是一个 Kind 的具体使用，比如 Pod 类型对应的资源是 pods。例如我们可以创建 5 个 pods 资源，其类型是 Pod。描述资源的单词都是小写的，就像 pods，而对应的类型一般就是这个资源的首字母大写单词的单数形式，比如 pods 对应 Pod。类型和资源往往是一一对应的，尤其是在 CRD 的实现上。常见的特例就是为了支持 HorizontalPodAutoscaler（HPA）和不同类型交互，Scale 类型对应的资源有 deployments/scale 和 replicasets/scale 两种。

于是我们知道了可以通过一个 GroupVersionKind（GVK）确定一个具体的类型，同样确定一个资源也就可以通过 GroupVersionResource（GVR）来实现。

3.4 本章小结

本章我们从"认识 Kubernetes API"开始，然后以几个实际操作示例来演示如何与 Kubernetes API 交互，最后介绍了非常重要的 GVK 的含义，希望大家通过本章的学习能够对 Kubernetes API 有一个更直观的认识。

第 4 章 理解 client-go

client-go项目就是用于和Kubernetes API Server通信的Go语言开发工具包。虽然使用Kubebuilder已经屏蔽了不少client-go的细节，但是要深入Operator开发机制，还是需要对client-go有一定的理解。本章先从使用的维度来介绍client-go提供的能力，下一章会深入client-go中相关模块的源码，让大家一览client-go的全貌。

4.1 client-go 项目介绍

我们先来认识client-go项目，查看client-go的项目结构、版本规则，以及如何获取client-go等。

4.1.1 client-go 的代码库

在GitHub（https://github.com/kubernetes/client-go）上可以找到client-go的源码，如图4-1所示。

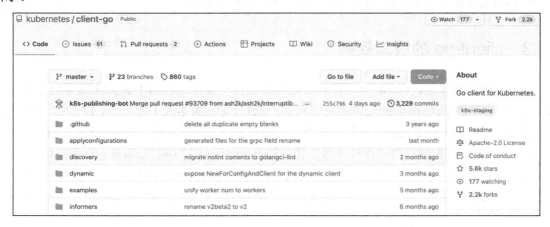

图 4-1 查看 client-go 源码

不过，这个库的代码是以每天一次的频率从kubernetes/kubernetes主库中自动同步过来的，所以如果大家想要给client-go提交PR，记得不能提交到这个代码库中。client-go在 kubernetes代码库中的位置如图4-2所示。

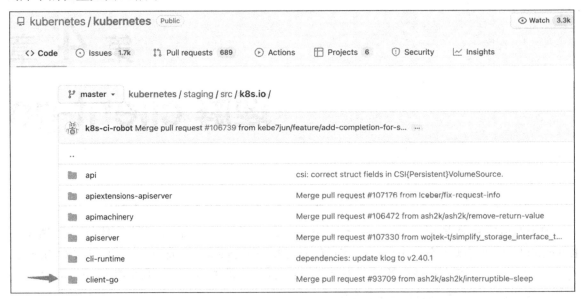

图 4-2　查看 client-go 的位置

4.1.2　client-go 的包结构

- kubernetes：这个包中放的是用client-gen自动生成的用来访问Kubernetes API的ClientSet，后面会经常看到ClientSet这个工具。
- discovery：这个包提供了一种机制用来发现API Server支持的API资源。
- dynamic：这个包中包含dynamic client，用来执行任意API资源对象的通用操作。
- plugin/pkg/client/auth：这个包提供了可选的用于获取外部源证书的认证插件。
- transport：这个包用于设置认证和建立连接。
- tools/cache：这个包中放了很多和开发控制器相关的工具集。

4.1.3　client-go 的版本规则

由于一些历史原因，client-go的版本规则经历了几次变化。我们不去关注很早的版本都有哪些规则，简单理解client-go的版本就是一句话：Kubernetes版本大于或等于1.17.0时，cllient-go版本使用对应的v0.x.y；Kubernetes版本小于1.17.0时，client-go版本使用kubernetes-1.x.y。其中，x和y与Kubernetes版本号后两位保持一致，比如Kubernetes v1.17.0对应client-go v0.17.0。

这里说的client-go的版本体现在tag上，我们在client-go的GitHub代码库的tag列表中可以直观地看到这些tag。表4-1展示了以Kubernetes 1.17.0版本为中点，client-go和Kubernetes的版本对应关系。

表 4-1 client-go 与 Kubernetes 的版本对应关系

client-go	Kubernetes				
	1.15.n	1.16.n	1.17.n	1.18.n	1.19.n
kubernetes-1.15.0	√				
kubernetes-1.16.0		√			
kubernetes-1.17.0/v0.17.0			√		
kubernetes-1.18.0/ v0.18.0				√	
kubernetes-1.19.0/ v0.19.0					√

如表4-1所示，第一行是Kubernetes版本，第一列是client-go版本。在Kubernetes 1.17.0版本之后，client-go老的版本号规则为了更好的兼容性还是保留着，不过最好还是使用新版本号v0.x.y这种格式。

另外，client-go代码库的分支规则和tag又稍有区别，下面简单地通过表4-2看一下Kubernetes 1.15.n版本之后两个代码库的分支规则对应关系。

表 4-2 Kubernetes 1.15.n 版本之后两个代码库的分支规则对应关系

client-go	Kubernetes
release-12.0	1.15
release-13.0	1.16
release-14.0	1.17
release-1.18	1.18
release-1.19	1.19

如表4-2所示，从1.18版本开始，两者的分支名称又对应起来了。其实client-go在Kubernetes 1.5版本以前就是现在的分支命名风格，不过从1.5之后变成了2.0，之后就是3.0、4.0、5.0……这种规则了，直到1.18版本。

4.1.4　获取 client-go

我们在写代码的时候需要使用client-go，第一步肯定是通过go get来获取相应版本的client-go依赖。如果需要新版本，可以直接执行：

```
go get k8s.io/client-go@latest
```

不过这样并不靠谱，我们一般需要选择明确的版本，最好是和自己使用的Kubernetes集群版本完全一致。本书使用v0.20.1版本的client-go，通过下面的命令来获取：

```
go get k8s.io/client-go@v0.20.1
```

4.2　client-go 使用示例

本节将通过3个示例来演示client-go的基础用法。

4.2.1　client-go 集群内认证配置

通过client-go访问API Server遇到的第一个问题是什么呢？当然是认证问题。前面说过API

Server是支持双向TLS认证的,换言之,我们随便编写一段代码发送HTTP请求给API Server肯定会因为认证问题而失败。下面通过client-go来编写一小段代码,完成认证后查询default命名空间下的Pod,并将其名字打印出来。我们将通过这个例子来入门client-go的使用。

1. 准备项目

```
# cd ~/MyOperatorProjects
# mkdir client-go-examples
# cd client-go-examples
# go mod init github.com/daniel-hutao/client-go-examples
go: creating new go.mod: module github.com/daniel-hutao/client-go-examples
# mkdir in-cluster-configuration
# cd in-cluster-configuration
# touch main.go
```

如上所示,我们创建一个新项目client-go-examples来存放接下来的几个示例程序。本小节的第一个示例将放在in-cluster-configuration目录下。还记得MyOperatorProjects目录吗?我们在前面提过,本书所有代码都会放到这个目录下。

2. 实现业务逻辑

代码如下:

```go
func main() {
    config, err := rest.InClusterConfig()
    if err != nil {
        log.Fatal(err)
    }

    clientset, err := kubernetes.NewForConfig(config)
    if err != nil {
        log.Fatal(err)
    }

    for {
        pods, err := clientset.CoreV1().Pods("default").
            List(context.TODO(), metav1.ListOptions{})
        if err != nil {
            log.Fatal(err)
        }
        log.Printf("There are %d pods in the cluster\n", len(pods.Items))
        for i, pod := range pods.Items {
            log.Printf("%d -> %s/%s", i+1, pod.Namespace, pod.Name)
        }
        <-time.Tick(5 * time.Second)
    }
}
```

这时import部分是这样的:

```
import (
    "context"
    "log"
    "time"

    metav1 "k8s.io/apimachinery/pkg/apis/meta/v1"
    "k8s.io/client-go/kubernetes"
    "k8s.io/client-go/rest"
)
```

另外，还需要关注一下依赖的版本，这段代码直接依赖的包有：

- k8s.io/apimachinery v0.20.1。
- k8s.io/client-go v0.20.1。

v0.20.1对应的Kubernetes版本也就是v1.20.1。接下来分析这段代码的逻辑。

步骤01 初始化 config：

```
config, err := rest.InClusterConfig()
if err != nil {
    log.Fatal(err)
}
```

在Kubernetes中，Pod创建时会自动把ServiceAccount token挂载到容器内的/var/run/secrets/kubernetes.io/serviceaccount路径下，这里的InClusterConfig()函数也是基于这个原理读取了认证所需的token和ca.crt两个文件。

步骤02 通过 config 初始化 clientset：

```
clientset, err := kubernetes.NewForConfig(config)
if err != nil {
    log.Fatal(err)
}
```

这里NewForConfig()函数返回了一个*Clientset对象，通过clientset可以实现各种资源的CURD操作。

步骤03 通过 clientset 来列出特定命名空间里的所有 Pod：

```
pods, err := clientset.CoreV1().Pods("default").
    List(context.TODO(), metav1.ListOptions{})
if err != nil {
    log.Fatal(err)
}
```

当然，这段代码我们放在一个for循环中，每次执行完都等待了5秒，不过这些都是一般逻辑，不再赘述。这里可以看到clientset提供了List Pod的能力，我们直接指定了命名空间是default，然后列出了里面的所有Pod。

步骤04 打印信息：

```
for i, pod := range pods.Items {
```

```
            log.Printf("%d -> %s/%s", i+1, pod.Namespace, pod.Name)
    }
```

这里的逻辑就比较简单了，前面List函数获取到的Pod是一个PodList类型的对象实例，PodList里有一个Items属性是[]Pod类型的，所以这里我们通过一个for-range循环来遍历这个Pod切片，然后打印出所有Pod的Namespace和Name。

3. 编写Dockerfile

```
FROM busybox
COPY ./app /app
ENTRYPOINT /app
```

我们只需简单地将代码放到容器中运行，所以Dockerfile不需要太复杂，够用就行。

4. 编译代码

```
cd in-cluster-configuration/
GOOS=linux go build -o ./in cluster .
```

5. 容器化并加载到kind环境

```
docker build -t in-cluster:v1 .
kind load docker-image in-cluster:v1 --name=dev
```

6. 创建ClusterRoleBinding

```
kubectl create clusterrolebinding default-view --clusterrole=view --serviceaccount=default:default
```

为了能够列出所有Pod，需要给default Service Account View权限。

7. 启动Pod

```
# kubectl run -i in-cluster --image=in-cluster:v1
2022/01/10 14:42:03 There are 1 pods in the cluster
2022/01/10 14:42:03 1 -> default/in-cluster
2022/01/10 14:42:08 There are 1 pods in the cluster
2022/01/10 14:42:08 1 -> default/in-cluster
2022/01/10 14:42:13 There are 1 pods in the cluster
2022/01/10 14:42:13 1 -> default/in-cluster
```

从日志中可以看到我们的代码运行起来了，而且查询到了default命名空间中的Pod，并将其信息打印了出来。在笔者的环境中，由于default命名空间中并没有其他Pod，因此这里in-cluster只查询到了自己，于是输出了default/in-cluster。

4.2.2 client-go 集群外认证配置

上一小节演示了client-go在集群内的认证方法，对应的集群外运行的服务又怎么解决认证问题进而访问API Server呢？我们已经知道了集群内Pod实现认证的核心原理是容器内自动挂载了token和ca.crt两个文件，那么容器外能使用的配置又是什么呢？其实也很简单，就像kubectl访问API Server一样，我们拿到kubeconfig文件后，就可以利用这个配置文件实现认证了。

1. 准备目录

```
cd ~/MyOperatorProjects
cd client-go-examples
mkdir out-of-cluster-configuration
cd out-of-cluster-configuration
touch main.go
```

这次我们还是在client-go-examples项目中实现这个示例，在in-cluster-configuration同级目录下创建out-of-cluster-configuration，然后在out-of-cluster-configuration目录中实现业务逻辑。

2. 实现业务逻辑

```go
func main() {
    homePath := homedir.HomeDir()
    if homePath == "" {
        log.Fatal("failed to get the home directory")
    }
    kubeconfig := filepath.Join(homePath, ".kube", "config")
    config, err := clientcmd.BuildConfigFromFlags("", kubeconfig)
    if err != nil {
        log.Fatal(err)
    }
    clientset, err := kubernetes.NewForConfig(config)
    if err != nil {
        log.Fatal(err)
    }
    for {
        pods, err := clientset.CoreV1().Pods("default").
            List(context.TODO(), metav1.ListOptions{})
        if err != nil {
            log.Fatal(err)
        }
        log.Printf("There are %d pods in the cluster\n", len(pods.Items))
        for i, pod := range pods.Items {
            log.Printf("%d -> %s/%s", i+1, pod.Namespace, pod.Name)
        }
        <-time.Tick(5 * time.Second)
    }
}
```

这时import部分的代码如下：

```go
import (
    "context"
    "log"
    "path/filepath"
    "time"
```

```
    metav1 "k8s.io/apimachinery/pkg/apis/meta/v1"
    "k8s.io/client-go/kubernetes"
    "k8s.io/client-go/tools/clientcmd"
    "k8s.io/client-go/util/homedir"
)
```

同样我们先来分析一下这段代码的逻辑，然后测试运行。这段代码与前面的in-cluster-configuration的主要区别是获取*restclient.Config的方式不同，后面config的使用是完全一致的。所以我们只需要关注config相关逻辑就可以了。

步骤 01 获取 kubeconfig 路径：

```
homePath:= homedir.HomeDir()
if homePath == "" {
    log.Fatal("failed to get the home directory")
}
kubeconfig:= filepath.Join(homePath, ".kube", "config")
```

这里的逻辑还是比较简单的，首先通过k8s.io/client-go/util/homedir包提供的HomeDir()函数获取用户家目录，然后拼接kubeconfig的地址。比如在笔者的MacBook上，kubeconfig的路径是/Users/danielhu/.kube/config。

步骤 02 通过 kubeconfig 初始化 config：

```
config, err := clientcmd.BuildConfigFromFlags("", kubeconfig)
if err != nil {
    log.Fatal(err)
}
```

BuildConfigFromFlags()函数的两个参数分别是string类型的masterUrl和kubeconfigPath。这里我们只给了第二个参数，因为kubeconfig中已经包含API Server的连接信息。BuildConfigFromFlags()会根据kubeconfig文件中的配置来初始化Config对象，然后和InClusterConfig()函数一样会返回一个*Config对象，拿到*Config后，我们就可以进一步初始化ClientSet，接下来的事情大家应该很熟悉了。

3. 编译运行

```
# cd out-of-cluster-configuration/
# go build -o out-of-cluster
# ./out-of-cluster
2022/01/11 22:57:44 There are 2 pods in the cluster
2022/01/11 22:57:44 1 -> default/demo
2022/01/11 22:57:44 2 -> default/hello-75f9fddff6-5mzdn
2022/01/11 22:57:49 There are 2 pods in the cluster
2022/01/11 22:57:49 1 -> default/demo
2022/01/11 22:57:49 2 -> default/hello-75f9fddff6-5mzdn
```

这次我们并没有去做容器化，因为直接在本地运行就已经足够说明问题了。可以看到在笔者当前的环境中存在两个Pod，分别是demo和hello-×××，大家在自己的环境中运行这个程序看到的输出结果肯定和笔者的不完全一致，不过效果是相同的。

4.2.3　client-go 操作 Deployment

前面我们已经学习了如何通过client-go调用Kubernetes API，相信大家已经迫不及待想要试一下如何用ClientSet来创建、更新、读取和删除具体的资源了。本小节就以Deployment资源的创建、更新、删除等操作为例演示一下相关操作流程。

1. 准备目录

```
cd ~/MyOperatorProjects
cd client-go-examples
mkdir handle-deployment
cd handle-deployment
touch main.go
```

这些步骤大家已经很熟悉了，这次将代码放到handle-deployment目录下。

2. 实现main函数

```go
func main() {
    homePath := homedir.HomeDir()
    if homePath == "" {
        log.Fatal("failed to get the home directory")
    }

    kubeconfig := filepath.Join(homePath, ".kube", "config")
    config, err := clientcmd.BuildConfigFromFlags("", kubeconfig)
    if err != nil {
        log.Fatal(err)
    }

    clientset, err := kubernetes.NewForConfig(config)
    if err != nil {
        log.Fatal(err)
    }

    dpClient := clientset.AppsV1().
        Deployments(corev1.NamespaceDefault)

    log.Println("create Deployment")
    if err := createDeployment(dpClient); err != nil {
        log.Fatal(err)
    }
    <-time.Tick(1 * time.Minute)

    log.Println("update Deployment")
    if err := updateDeployment(dpClient); err != nil {
        log.Fatal(err)
    }
    <-time.Tick(1 * time.Minute)

    log.Println("delete Deployment")
    if err := deleteDeployment(dpClient); err != nil {
        log.Fatal(err)
    }
}
```

```
        <-time.Tick(1 * time.Minute)
        log.Println("end")
}
```

这段代码不难理解,前面从config到clientset的逻辑和前面几个示例是一样的。后面的dpClient是通过clientset.AppsV1().Deployments(corev1.NamespaceDefault)调用获得的,这是一个DeploymentInterface类型,可以用来对Deployment类型进行各种操作,比如Get()、List()、Watch()、Create()、Update()、Delete()、Patch()等。

接着就是create、update、delete三个函数的调用,间隔是1分钟。我们通过这种方式让代码等待一会儿,好有时间观察集群状态。接下来自然就是这三个函数的实现了。

3. 实现createDeployment()函数

```
func createDeployment(dpClient v1.DeploymentInterface) error {
    replicas := int32(3)
    newDp := &appsv1.Deployment{
        ObjectMeta: metav1.ObjectMeta{
            Name: "nginx-deploy",
        },
        Spec: appsv1.DeploymentSpec{
            Replicas: &replicas,
            Selector: &metav1.LabelSelector{
                MatchLabels: map[string]string{
                    "app": "nginx",
                },
            },
            Template: corev1.PodTemplateSpec{
                ObjectMeta: metav1.ObjectMeta{
                    Labels: map[string]string{
                        "app": "nginx",
                    },
                },
                Spec: corev1.PodSpec{
                    Containers: []corev1.Container{
                        {
                            Name:  "nginx",
                            Image: "nginx:1.14",
                            Ports: []corev1.ContainerPort{
                                {
                                    Name:          "http",
                                    Protocol:      corev1.ProtocolTCP,
                                    ContainerPort: 80,
                                },
                            },
                        },
                    },
                },
            },
        },
    }
```

```
    _, err := dpClient.Create(context.TODO(),
        newDp, metav1.CreateOptions{})
    return err
}
```

这段代码看着不短,其实逻辑非常简单。我们把Deployment对象的赋值折叠起来,剩下的逻辑就只有几行了:

```
replicas := int32(3)
newDp := &appsv1.Deployment{...}
_, err := dpClient.Create(context.TODO(),
    newDp, metav1.CreateOptions{})
return err
```

创建一个Deployment就是这几行,先构造一个newDp,然后调用dpClient的Create()函数来创建这个Deployment。newDp中的配置内容和前面用过的nginx-deploy是完全一样的,这里镜像版本设置了nginx:1.14,副本数是3,等一下更新操作时会将这个镜像改为nginx:1.16。

4. 实现updateDeployment()函数

```
func updateDeployment(dpClient v1.DeploymentInterface) error {
    dp, err := dpClient.Get(context.TODO(),
        "nginx-deploy", metav1.GetOptions{})
    if err != nil {
        return err
    }
    dp.Spec.Template.Spec.Containers[0].Image = "nginx:1.16"
    return retry.RetryOnConflict(
        retry.DefaultRetry, func() error {
            _, err = dpClient.Update(context.TODO(),
                dp, metav1.UpdateOptions{})
            return err
        },
    )
}
```

update的逻辑主要分三步:

步骤01 获取 nginx-deploy:

```
    dp, err := dpClient.Get(context.TODO(),
        "nginx-deploy", metav1.GetOptions{})
    if err != nil {
        return err
    }
```

我们通过dpClient的Get()方法获取到一个*Deployment对象,然后就可以操作这个对象了。

步骤02 修改镜像字段:

```
    dp.Spec.Template.Spec.Containers[0].Image = "nginx:1.16"
```

我们把Image配置成nginx:1.16,这样会触发一次Pod的滚动更新。

步骤 03 调用 Update() 函数完成更新：

```
return retry.RetryOnConflict(
    retry.DefaultRetry, func() error {
        _, err = dpClient.Update(context.TODO(),
            dp, metav1.UpdateOptions{})
        return err
    },
)
```

这里的主要逻辑调用dpClient.Update()来完成Deployment的更新。外层包的RetryOnConflict()函数只是一种健壮性手段，如果Update过程失败了，这里能提供重试机制。RetryOnConflict()函数签名是这样的：

```
RetryOnConflict(backoff wait.Backoff, fn func() error) error
```

5. 实现deleteDeployment()函数

```
func deleteDeployment(dpClient v1.DeploymentInterface) error {
    deletePolicy := metav1.DeletePropagationForeground
    return dpClient.Delete(
        context.TODO(), "nginx-deploy", metav1.DeleteOptions{
            PropagationPolicy: &deletePolicy,
        },
    )
}
```

看过了前面的创建和更新Deployment操作后，这里的删除应该就很简单了。这里同样通过dpClient的Delete()方法完成删除操作，里面的PropagationPolicy属性配置了metav1.DeletePropagationForeground，这里有三种可选特性：

- DeletePropagationOrphan：不考虑依赖资源。
- DeletePropagationBackground：后台删除依赖资源。
- DeletePropagationForeground：前台删除依赖资源。

6. 测试运行

```
# cd handle-deployment/
# go run main.go
2022/01/12 22:14:29 create Deployment
2022/01/12 22:15:29 update Deployment
2022/01/12 22:16:29 delete Deployment
2022/01/12 22:17:29 end
```

日志很简洁，只是打印出来了几个关键步骤，并且留了1分钟间隔，用来供我们观察现象。在这个程序运行的过程中，我们可以新开一个窗口观察资源的变化，大致可以看到如下信息：

1）Deployment 的创建：

```
# kubectl get deployment
NAME           READY   UP-TO-DATE   AVAILABLE   AGE
nginx-deploy   3/3     3            3           12s
```

这时可以看到镜像版本是什么：

```
# kubectl describe deployment nginx-deploy | grep Image
    Image:        nginx:1.14
```

2）Deployment 的更新：

```
# kubectl get pod
NAME                              READY   STATUS              RESTARTS   AGE
nginx-deploy-6d96cc64b7-9kz8r     0/1     ContainerCreating   0          13s
nginx-deploy-94f57779c-8jzb4      1/1     Running             0          73s
nginx-deploy-94f57779c-gtbvc      1/1     Running             0          73s
nginx-deploy-94f57779c-xcrkp      1/1     Running             0          73s
```

由于需要下载新镜像，因此第一个新Pod的创建耗时会久一些。当第一个Pod起来后，另外几个新Pod的创建和旧Pod的删除会很快：

```
# kubectl get pod
NAME                              READY   STATUS    RESTARTS   AGE
nginx-deploy-6d96cc64b7-48x25     1/1     Running   0          7s
nginx-deploy-6d96cc64b7-9kz8r     1/1     Running   0          56s
nginx-deploy-6d96cc64b7-xxb54     1/1     Running   0          6s
```

3）Deployment 的删除：

```
# kubectl get deployment
No resources found in default namespace.
```

4.3 本章小结

本章我们从client-go项目的代码库、项目包结构等基础知识入手，进而演示了使用client-go来完成认证、操作Deployment等过程。希望通过本章的学习，大家能够对如何使用client-go有一个直观的认识。

第 5 章
client-go 源码分析

源码分析部分有一定难度，如果大家感觉阅读起来比较困难，可以先跳到下一章学习其他开发技巧，等到有一定的实践经验，对Operator开发流程比较熟悉后，再回过头来学习client-go的源码，从而对Operator的原理和能力有更深的认识。

我们在深度使用Kubernetes时难免会涉及Operator的开发，目前虽然已经有Kubebuilder/Operator SDK、controller-runtime等工具可以较好地屏蔽底层细节，让我们专注于自身的业务逻辑，但是不清楚底层原理会让我们在编码过程中心里没底，比如自定义控制器重启时是否会重新收到所有相关Event，调谐的子资源是Deployment时相关Pod的变更是否会触发调谐逻辑等，很多细节问题会不停地跳出来，让你对自己的代码没有信心。所以我们只有详细分析过client-go和Operator开发相关的各种组件的原理与源码后，才能对自己开发的自定义控制器行为知根知底，胸有成竹。

5.1 client-go 源码概览

为了尽量避免本书出版后大家看到的源码版本落后太多，笔者在2021年提前刷完了相关模块代码，并且整理了相关笔记，实时发布在了笔者的个人公众号"胡说云原生"上，现在再次更新本地代码到新版本，然后参考之前的笔记，重新整理一版"源码分析笔记"，这样大家看到本书时，自己在工作中使用的Kubernetes版本会和本章内容所基于的源码版本相差很小。

5.1.1 关于 client-go 源码版本

现在是2022年2月3日23时，笔者在Kubernetes本地代码库根目录下执行相应git命令来更新代码：

```
# git fetch upstream
remote: Enumerating objects: 25134, done.
```

```
remote: Counting objects: 100% (11749/11749), done.
remote: Compressing objects: 100% (32/32), done.
remote: Total 25134 (delta 11717), reused 11748 (delta 11717), pack-reused 13385
Receiving objects: 100% (25134/25134), 23.83 MiB | 171.00 KiB/s, done.
Resolving deltas: 100% (18426/18426), completed with 3602 local objects.
From https://github.com/kubernetes/kubernetes
   87b0412232d..adbda068c18  master          -> upstream/master
```

然后rebase本地master分支：

```
# git rebase upstream/master
First, rewinding head to replay your work on top of it...
Fast-forwarded master to upstream/master.
```

这时本地分支状态是：

- Commit：adbda068c1808fcc8a64a94269e0766b5c46ec41。
- Date：Thu Feb 3 04:31:44 2022 -0800。
- Branch：master。

也就是说，接下来的源码分析内容都会基于master分支的"2022年2月3日"更新的代码来展开。

由于client-go是staging到Kubernetes项目中的，这里直接更新了Kubernetes项目，而不是下载单独的client-go项目，因此在阅读client-go源码时，为了避免目录结构嵌套过深，我们可以直接打开staging/k8s.io/client-go目录，如图5-1所示。

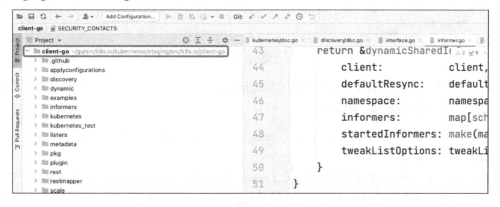

图 5-1　打开 client-go 目录

5.1.2　client-go 模块概览

client-go项目是与kube-apiserver通信的clients的具体实现，其中包含很多相关工具包，例如kubernetes包就包含与Kubernetes API通信的各种ClientSet，而tools/cache包则包含很多强大的编写控制器相关的组件。所以接下来我们会以自定义控制器的底层实现原理为线索，来分析client-go中相关模块的源码实现。

如图5-2所示，我们在编写自定义控制器的过程中大致依赖于如下组件，其中圆形的是自定义控制器中需要编码的部分，其他椭圆和圆角矩形的是client-go提供的一些"工具"。

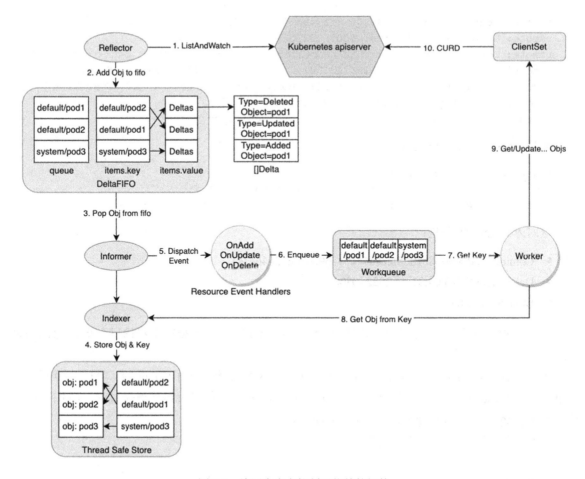

图 5-2 编写自定义控制器依赖的组件

client-go的源码入口在Kubernetes项目的staging/src/k8s.io/client-go中,先整体查看上面涉及的相关模块,然后逐个深入分析其实现。

- Reflector:Reflector从apiserver监听(watch)特定类型的资源,拿到变更通知后,将其丢到DeltaFIFO队列中。
- Informer:Informer从DeltaFIFO中弹出(pop)相应对象,然后通过Indexer将对象和索引丢到本地cache中,再触发相应的事件处理函数(Resource Event Handlers)。
- Indexer:Indexer主要提供一个对象根据一定条件检索的能力,典型的实现是通过namespace/name来构造key,通过Thread Safe Store来存储对象。
- WorkQueue:WorkQueue一般使用的是延时队列实现,在Resource Event Handlers中会完成将对象的key放入WorkQueue的过程,然后在自己的逻辑代码里从WorkQueue中消费这些key。
- ClientSet:ClientSet提供的是资源的CURD能力,与apiserver交互。
- Resource Event Handlers:我们一般在Resource Event Handlers中添加一些简单的过滤功能,判断哪些对象需要加到WorkQueue中进一步处理,对于需要加到WorkQueue中的对象,就提取其key,然后入队。
- Worker:Worker指的是我们自己的业务代码处理过程,在这里可以直接收到WorkQueue中的任务,可以通过Indexer从本地缓存检索对象,通过ClientSet实现对象的增、删、改、查逻辑。

5.2 WorkQueue 源码分析

我们前面提到了WorkQueue一般使用延时队列实现，在Resource Event Handlers中会完成将对象的key放入WorkQueue的过程，然后在自己的逻辑代码里从WorkQueue中消费这些key。

client-go的util/workqueue包里主要有三个队列，分别是普通队列、延时队列和限速队列，后一个队列以前一个队列的实现为基础，层层添加新功能，我们按照Queue、DelayingQueue、RateLimitingQueue的顺序层层拨开来看各种队列是如何实现的。

在k8s.io/client-go/util/workqueue包下可以看到这样三个Go源文件：

- queue.go。
- delaying_queue.go。
- rate_limiting_queue.go。

很明显这三个文件分别对应三种队列实现，下面我们逐个对它们进行分析。

5.2.1 普通队列 Queue 的实现

1. 表示Queue的接口和相应的实现结构体

定义Queue的接口在queue.go中直接叫作Interface，代码如下：

```
type Interface interface {
    Add(item interface{})              // 添加一个元素
    Len() int                          // 元素个数
    Get() (item interface{}, shutdown bool) // 获取一个元素，第二个返回值和channel类
似，标记队列是否关闭了
    Done(item interface{})             // 标记一个元素已经处理完
    ShutDown()                         // 关闭队列
    ShutDownWithDrain()                // 关闭队列，但是等待队列中元素处理完
    ShuttingDown() bool                // 标记当前channel是否正在关闭
}
```

从接口定义中，我们可以很清晰地看到Queue提供了哪些能力，接着我们继续看是哪个结构体实现了这个接口。Interface的实现类型是Type，这个名字延续了用Interface表示interface的风格，源码还是在queue.go中：

```
type Type struct {
    queue []t          // 定义元素的处理顺序，里面所有元素在dirty集合中应该都有，而不能出现在
processing集合中
    dirty set          // 标记所有需要被处理的元素
    processing set     // 当前正在被处理的元素，当处理完后，需要检查该元素是否在dirty集合中，
如果在则添加到queue队列中

    cond *sync.Cond
    shuttingDown bool
    drain        bool
```

```
metrics           queueMetrics
unfinishedWorkUpdatePeriod time.Duration
clock                      clock.WithTicker
}
```

这个Queue的工作逻辑大致是这样的，里面的三个属性queue、dirty、processing都保存有元素（items），但是含义有所不同：

- queue：这是一个[ ]t类型，也就是一个切片，因为其有序，所以这里当作一个列表来存储元素的处理顺序。
- dirty：属于set类型，dirty就是一个集合，其中存储的是所有需要处理的元素，这些元素也会保存在queue中，但是集合中的元素是无序的，且集合的特性是其里面的元素具有唯一性。
- processing：也是一个集合，存放的是当前正在处理的元素，也就是说这个元素来自queue出队的元素，同时这个元素会被从dirty中删除。

紧接着大家好奇的问题肯定是这里的set类型是怎么定义的：

```
type empty struct{}
type t interface{}
type set map[t]empty

func (s set) has(item t) bool {
    _, exists := s[item]
    return exists
}

func (s set) insert(item t) {
    s[item] = empty{}
}

func (s set) delete(item t) {
    delete(s, item)
}

func (s set) len() int {
    return len(s)
}
```

可以看到上述的set其实是一个map（映射），利用map key的唯一性来当作set使用，这里的key是t类型，和属性queue的类型[ ]t中的t是同一个类型。set类型实现了has()、insert()、delete()、len()几个方法，用于支持集合类型的基本操作。

我们继续来看Type结构体的几个主要方法的实现。

2. Queue.Add()方法的实现

Add()方法用于标记一个新的元素需要被处理，代码也在queue.go内，如下所示：

```
func (q *Type) Add(item interface{}) {
    q.cond.L.Lock()
    defer q.cond.L.Unlock()
    if q.shuttingDown {                    // 如果queue正在被关闭，则返回
        return
```

```go
    }
    if q.dirty.has(item) {                    // 如果dirty set中已经有了该元素，则返回
        return
    }

    q.metrics.add(item)

    q.dirty.insert(item)                      // 添加到dirty set 中
    if q.processing.has(item) {               // 如果正在被处理，则返回
        return
    }
    // 如果没有正在处理，则加到q.queue中
    q.queue = append(q.queue, item)
    q.cond.Signal()。                         //通知getter有新元素到来
}
```

3. Queue.Get()方法的实现

Get()方法在获取不到元素的时候会阻塞，直到有一个元素可以被返回。这个方法同样在queue.go文件中实现，我们具体来看：

```go
func (q *Type) Get() (item interface{}, shutdown bool) {
    q.cond.L.Lock()
    defer q.cond.L.Unlock()
    // 如果q.queue为空，并且没有正在关闭，则等待下一个元素的到来
    for len(q.queue) == 0 && !q.shuttingDown {
        q.cond.Wait()
    }
    // 这时如果q.queue长度还是0，则说明q.shuttingDown为true，所以直接返回
    if len(q.queue) == 0 {
        return nil, true
    }

    item = q.queue[0]           // 获取q.queue第一个元素
    q.queue[0] = nil  // 这里的nil赋值是为了让底层数组不再引用元素对象，从而能够被垃圾回收
    q.queue = q.queue[1:] // 更新q.queue

    q.metrics.get(item)
    // 刚才获取到的q.queue第一个元素放到processing集合中
    q.processing.insert(item)
    q.dirty.delete(item)        // 在dirty集合中删除该元素

    return item, false          // 返回元素
}
```

4. Queue.Done()方法的实现

Done()方法的作用是标记一个元素已经处理完成，代码还是在queue.go中：

```go
func (q *Type) Done(item interface{}) {
    q.cond.L.Lock()
    defer q.cond.L.Unlock()

    q.metrics.done(item)
    // 在processing集合中删除该元素
    q.processing.delete(item)
```

```
        if q.dirty.has(item) { // 如果dirty中还有，则说明还需要再次处理，放到q.queue中
            q.queue = append(q.queue, item)
            q.cond.Signal()// 通知getter有新的元素
        } else if q.processing.len() == 0 {
            q.cond.Signal()
        }
    }
```

5.2.2 延时队列 DelayingQueue 的实现

1. 表示DelayingQueue的接口和相应的实现结构体

定义DelayingQueue的接口在delaying_queue.go源文件中，名字和Queue所使用的Interface很对称，叫作DelayingInterface：

```
type DelayingInterface interface {
    Interface
    AddAfter(item interface{}, duration time.Duration)
}
```

可以看到DelayingInterface接口中嵌套了一个表示Queue的Interface，也就是说DelayingInterface接口包含Interface接口的所有方法声明。另外相比于Queue，这里只是多了一个AddAfter()方法，顾名思义，也就是延时添加元素的意思。

我们继续看对应实现这个接口的结构体：

```
type delayingType struct {
    Interface                          // 嵌套普通队列 Queue
    clock clock.Clock                  // 计时器
    stopCh chan struct{}
    stopOnce sync.Once                 // 用来确保ShutDown()方法只执行一次
    // 默认10秒的心跳，后用在一个大循环里，避免没有新元素时一直阻塞
    heartbeat clock.Ticker
    // 传递waitFor的channel，默认大小为1000
    waitingForAddCh chan *waitFor
    metrics retryMetrics
}
```

对于延时队列，我们关注的入口方法肯定就是新增的AddAfter()，在查看这个方法的具体逻辑前，先了解上面delayingType中涉及的waitFor类型。

2. waitFor对象

waitFor的实现也在delaying_queue.go中，其结构体定义如下：

```
type waitFor struct {
    data    t                  // 准备添加到队列中的数据
    readyAt time.Time          // 应该被加入队列的时间
    index   int                // 在heap中的索引
}
```

在waitFor结构体下面有这样一行代码：

```go
type waitForPriorityQueue []*waitFor
```

这里定义了一个waitFor的优先级队列，用最小堆的方式来实现，这个类型实现了heap.Interface接口。我们具体看一下源码：

```go
// 添加一个元素到队列中
func (pq *waitForPriorityQueue) Push(x interface{}) {
    n := len(*pq)
    item := x.(*waitFor)
    item.index = n
    *pq = append(*pq, item)
}

// 从队列尾巴移除一个元素
func (pq *waitForPriorityQueue) Pop() interface{} {
    n := len(*pq)
    item := (*pq)[n-1]
    item.index = -1
    *pq = (*pq)[0:(n - 1)]
    return item
}

// 获取队列第一个元素
func (pq waitForPriorityQueue) Peek() interface{} {
    return pq[0]
}
```

3. NewDelayingQueue

在分析AddAfter()方法的实现之前，我们先了解DelayingQueue的几个New函数：

```go
// 这个函数没有多少逻辑，只是接收一个Name，然后继续调用另一个New函数
func NewNamedDelayingQueue(name string) DelayingInterface {
    return NewDelayingQueueWithCustomClock(clock.RealClock{}, name)
}

// 在前一个New函数的基础上，这里增加了一个指定clock的能力
func NewDelayingQueueWithCustomClock(clock clock.WithTicker, name string) DelayingInterface {
    // 注意这里的NewNamed()函数调用，后面还会分析到
    return newDelayingQueue(clock, NewNamed(name), name)
}

func newDelayingQueue(clock clock.WithTicker, q Interface, name string) *delayingType {
    ret := &delayingType{
        Interface:        q,
        clock:            clock,
        heartbeat:        clock.NewTicker(maxWait), // 10秒一次心跳
        stopCh:           make(chan struct{}),
        waitingForAddCh:  make(chan *waitFor, 1000),
        metrics:          newRetryMetrics(name),
    }
```

```
        go ret.waitingLoop()        // 留意这里的函数调用，后面还会分析到
        return ret
}
```

上面在注释中提到了两处需要注意的细节：

- NewNamed()函数。
- waitingLoop()方法。

NewNamed()函数用于创建一个前面提到的Queue的对应类型Type对象，这个值被传递给了newDelayingQueue()函数，进而赋值给了delayingType{}对象的Interface字段，于是后面delayingType类型才能直接调用Type类型实现的方法。NewNamed()函数实现在queue.go中，代码如下：

```
func NewNamed(name string) *Type {
    rc := clock.RealClock{}
    return newQueue(
        rc,
        globalMetricsFactory.newQueueMetrics(name, rc),
        defaultUnfinishedWorkUpdatePeriod,
    )
}
```

另一个细节是waitingLoop()方法，这里的逻辑有点复杂，我们接下来单独分析。

4. waitingLoop()方法

waitingLoop()方法是延时队列实现的核心逻辑所在。

```
func (q *delayingType) waitingLoop() {
    defer utilruntime.HandleCrash()

    // 队列里没有元素时实现等待
    never := make(<-chan time.Time)
    var nextReadyAtTimer clock.Timer
    // 构造一个优先级队列
    waitingForQueue := &waitForPriorityQueue{}
    // 这一行其实功能上没有什么作用，不过在可读性上有点帮助
    heap.Init(waitingForQueue)
    // 这个map用来处理重复添加逻辑，下面会讲到
    waitingEntryByData := map[t]*waitFor{}
    // 无限循环
    for {
        // 这个地方Interface从语法上来看可有可无，不过放在这里能够强调调用了内部Queue的ShuttingDown()方法
        if q.Interface.ShuttingDown() {
            return
        }

        now := q.clock.Now()
        // 队列里有元素就开始循环
```

```go
for waitingForQueue.Len() > 0 {
    // 获取第一个元素
    entry := waitingForQueue.Peek().(*waitFor)
    if entry.readyAt.After(now) { // 时间还没到，先不处理
        break
    }
    // 时间到了，pop出第一个元素。注意waitingForQueue.Pop()是最后一个元素，
    // heap.Pop()是第一个元素
    entry = heap.Pop(waitingForQueue).(*waitFor)
    // 将数据加到延时队列里
    q.Add(entry.data)
    // 在map中删除已经加到延时队列的元素
    delete(waitingEntryByData, entry.data)
}
// 如果队列中有元素，就用第一个元素的等待时间初始化计时器，如果为空则一直等待
nextReadyAt := never
if waitingForQueue.Len() > 0 {
    if nextReadyAtTimer != nil {
        nextReadyAtTimer.Stop()
    }
    entry := waitingForQueue.Peek().(*waitFor)
    nextReadyAtTimer = q.clock.NewTimer(entry.readyAt.Sub(now))
    nextReadyAt = nextReadyAtTimer.C()
}

select {
case <-q.stopCh:
    return

case <-q.heartbeat.C():   // 心跳时间是10秒，到了就继续下一轮循环
case <-nextReadyAt:       // 第一个元素的等待时间到了，继续下一轮循环
// waitingForAddCh收到新的元素
case waitEntry := <-q.waitingForAddCh:
    // 如果时间没到，就加到优先级队列里；如果时间到了，就直接加到延时队列里
    if waitEntry.readyAt.After(q.clock.Now()) {
        insert(waitingForQueue, waitingEntryByData, waitEntry)
    } else {
        q.Add(waitEntry.data)
    }
    // 下面的逻辑就是将waitingForAddCh中的数据处理完
    drained := false
    for !drained {
        select {
        case waitEntry := <-q.waitingForAddCh:
            if waitEntry.readyAt.After(q.clock.Now()) {
                insert(waitingForQueue, waitingEntryByData, waitEntry)
            } else {
                q.Add(waitEntry.data)
            }
```

```go
            default:
                drained = true
            }
        }
    }
}
```

这个方法还有一个insert()函数调用，我们再来研究插入函数的逻辑：

```go
func insert(q *waitForPriorityQueue, knownEntries map[t]*waitFor, entry *waitFor) {
    // 这里的主要逻辑是看一个entry（表项）是否存在，如果已经存在，则新的entry的就绪时间更短，就更新时间
    existing, exists := knownEntries[entry.data]
    if exists {
        if existing.readyAt.After(entry.readyAt) {
            existing.readyAt = entry.readyAt // 如果存在就只更新时间
            heap.Fix(q, existing.index)
        }
        return
    }
    // 如果不存在就丢到q里，同时在map中记录一下，用于查重
    heap.Push(q, entry)
    knownEntries[entry.data] = entry
}
```

接下来终于可以看AddAfter()方法的实现了。

5. AddAfter()方法

AddAfter()方法的作用是在指定的延时时长到达之后，在work queue中添加一个元素。有了前面的铺垫后，这个方法的实现就很简单了，源码依旧在delaying_queue.go文件中：

```go
func (q *delayingType) AddAfter(item interface{}, duration time.Duration) {
    if q.ShuttingDown() { // 已经在关闭中就直接返回
        return
    }

    q.metrics.retry()

    if duration <= 0 { // 如果时间到了，就直接添加
        q.Add(item)
        return
    }

    select {
    case <-q.stopCh:
    // 构造waitFor{}，丢到waitingForAddCh
    case q.waitingForAddCh <- &waitFor{data: item, readyAt: q.clock.Now().Add(duration)}:
    }
}
```

5.2.3 限速队列 RateLimitingQueue 的实现

1. 表示RateLimitingQueue的接口和相应的实现结构体

不难猜到RateLimitingQueue对应的接口肯定叫作RateLimitingInterface，源码当然是在rate_limiting_queue.go中：

```
type RateLimitingInterface interface {
    DelayingInterface // 和延时队列里内嵌了普通队列一样，限速队列中内嵌了延时队列
    AddRateLimited(item interface{})        // 限速方式往队列中加入一个元素
    Forget(item interface{})                // 标识一个元素结束重试
    NumRequeues(item interface{}) int       // 标识这个元素被处理多少次了
}
```

实现RateLimitingInterface的结构体是rateLimitingType：

```
type rateLimitingType struct {
    DelayingInterface

    rateLimiter RateLimiter
}
```

这里出现了一个叫作RateLimiter的限速器，在看这个限速器之前，我们先分析这里的几个New函数。

2. RateLimitingQueue的New函数

```
func NewRateLimitingQueue(rateLimiter RateLimiter) RateLimitingInterface {
    return &rateLimitingType{
        DelayingInterface: NewDelayingQueue(),
        rateLimiter:       rateLimiter,
    }
}

func NewNamedRateLimitingQueue(rateLimiter RateLimiter, name string) RateLimitingInterface {
    return &rateLimitingType{
        DelayingInterface: NewNamedDelayingQueue(name),
        rateLimiter:       rateLimiter,
    }
}
```

两个函数的区别就是里面的延时队列有没有指定的名字，可以看到这里的逻辑非常简短，都需要一个限速器rateLimiter，然后就是调用DelayingQueue的几个New函数来填充内部的DelayingQueue。不难猜到这里的限速器会是一个核心对象，我们后面继续来看限速器是怎么实现的。

3. RateLimiter

RateLimiter表示一个限速器，我们看一下限速器是什么意思。RateLimiter定义在同一个包的default_rate_limiters.go源文件中，接口代码如下：

```
type RateLimiter interface {
```

```
            // 返回一个元素需要等待的时长
            When(item interface{}) time.Duration
            Forget(item interface{})   // 标识一个元素结束重试
            NumRequeues(item interface{}) int // 标识这个元素被处理多少次了
}
```

这个接口有5个实现，分别是：

- BucketRateLimiter。
- ItemExponentialFailureRateLimiter。
- ItemFastSlowRateLimiter。
- MaxOfRateLimiter。
- WithMaxWaitRateLimiter。

BucketRateLimiter 用了 Go 语言标准库的 golang.org/x/time/rate.Limiter 包实现。BucketRateLimiter实例化的时候，比如传递一个rate.NewLimiter(rate.Limit(10),100)进去，表示令牌桶里最多有100个令牌，每秒发放10个令牌。

```
type BucketRateLimiter struct {
    *rate.Limiter
}

var _ RateLimiter = &BucketRateLimiter{}

func (r *BucketRateLimiter) When(item interface{}) time.Duration {
    // 过多久后给当前元素发放一个令牌
    return r.Limiter.Reserve().Delay()
}

func (r *BucketRateLimiter) NumRequeues(item interface{}) int {
    return 0
}

func (r *BucketRateLimiter) Forget(item interface{}) {
}
```

ItemExponentialFailureRateLimiter这个限速器从名字上大概就能猜到是失败次数越多，限速越长，而且是呈指数级增长的一种限速器。ItemExponentialFailureRateLimiter的属性很简单，基本可以望文生义：

```
type ItemExponentialFailureRateLimiter struct {
    failuresLock sync.Mutex
    failures     map[interface{}]int

    baseDelay time.Duration
    maxDelay  time.Duration
}
```

核心逻辑是When()方法实现的：

```
func (r *ItemExponentialFailureRateLimiter) When(item interface{}) time.Duration {
    r.failuresLock.Lock()
    defer r.failuresLock.Unlock()
```

```
    exp := r.failures[item]
    r.failures[item] = r.failures[item] + 1    // 失败次数加1
    // 每调用一次，exp也就加1，对应到这里时2^n指数爆炸
    backoff := float64(r.baseDelay.Nanoseconds()) * math.Pow(2, float64(exp))
    // 如果超过了最大整型，就返回最大延时，不然后面的时间转换就会溢出
    if backoff > math.MaxInt64 {
        return r.maxDelay    // 如果超过最大延时，则返回最大延时
    }

    calculated := time.Duration(backoff)
    if calculated > r.maxDelay {
        return r.maxDelay
    }

    return calculated
}
```

剩下的 NumRequeues() 方法和 Forget() 方法太简单了，此处不再赘述。

ItemFastSlowRateLimiter 顾名思义也就是"快慢限速器"的意思，快慢指的是定义一个阈值，达到阈值之前快速重试，超过了就慢慢重试。这个结构体也不复杂：

```
type ItemFastSlowRateLimiter struct {
    failuresLock sync.Mutex
    failures     map[interface{}]int

    maxFastAttempts int                 // 快速重试的次数
    fastDelay       time.Duration       // 快重试间隔
    slowDelay       time.Duration       // 慢重试间隔
}
```

它的 When() 方法是这样实现的：

```
func (r *ItemFastSlowRateLimiter) When(item interface{}) time.Duration {
    r.failuresLock.Lock()
    defer r.failuresLock.Unlock()

    r.failures[item] = r.failures[item] + 1          // 标识重试次数 + 1
    // 如果快重试次数没有用完，则返回fastDelay
    if r.failures[item] <= r.maxFastAttempts {
        return r.fastDelay
    }
    // 反之返回slowDelay
    return r.slowDelay
}
```

至于 MaxOfRateLimiter，这个限速器是通过维护多个限速器列表，然后返回其中限速最严格的一个延时：

```
type MaxOfRateLimiter struct {
    limiters []RateLimiter
}

func (r *MaxOfRateLimiter) When(item interface{}) time.Duration {
    ret := time.Duration(0)
```

```
    for _, limiter := range r.limiters {
        curr := limiter.When(item)
        if curr > ret {
            ret = curr
        }
    }
    return ret
}
```

最后来看WithMaxWaitRateLimiter，这个限速器也很简单，就是在其他限速器上包装一个最大延迟的属性，如果到了最大延时，则直接返回：

```
type WithMaxWaitRateLimiter struct {
    limiter  RateLimiter
    maxDelay time.Duration
}
```

When()方法：

```
func (w WithMaxWaitRateLimiter) When(item interface{}) time.Duration {
    delay := w.limiter.When(item)
    if delay > w.maxDelay {
        return w.maxDelay
    }

    return delay
}
```

有了限速器的相关知识，想必接着看限速队列RateLimitingQueue的实现就很简单了。

4. RateLimitingQueue的限速实现

```
func (q *rateLimitingType) AddRateLimited(item interface{}) {
    // 内部存了一个延时队列，通过限速器计算出一个等待时间，然后传给延时队列
    q.DelayingInterface.AddAfter(item, q.rateLimiter.When(item))
}

func (q *rateLimitingType) NumRequeues(item interface{}) int {
    return q.rateLimiter.NumRequeues(item)
}

func (q *rateLimitingType) Forget(item interface{}) {
    q.rateLimiter.Forget(item)
}
```

可以看到限速队列的实现基本由内部的延时队列提供的功能和包装的限速器提供的功能组合而来，这里的代码就没有太多额外的逻辑了。

5.2.4 小结

我们在开发自定义控制器的时候，用到的WorkQueue就是使用这里的延时队列实现的，在Resource Event Handlers中会完成将对象的key（键）放入WorkQueue的过程，然后我们在自己的

逻辑代码里从WorkQueue中消费这些key。一个延时队列也就是实现了item（元素）的延时入队效果，内部是一个"优先级队列"，用了"最小堆"（有序完全二叉树），所以"在requeueAfter中指定一个调谐过程1分钟后重试"的实现原理也就清晰了。

5.3 DeltaFIFO 源码分析

DeltaFIFO也是一个重要组件，我们来继续研究client-go中DeltaFIFO的相关实现。DeltaFIFO相关代码在k8s.io/client-go/tools/cache包中，下文在不明确提到其他包名的情况下，所有源文件都指的是这个包内的源文件。

5.3.1 Queue 接口与 DeltaFIFO 的实现

1. Queue和Store接口

在fifo.go中定义了一个Queue接口，本节的主角DeltaFIFO就是Queue接口的一个实现。我们先来看Queue接口的代码：

```
type Queue interface {
    Store
    Pop(PopProcessFunc) (interface{}, error) // 会阻塞，直到有一个元素可以被pop出来，或者队列关闭
    AddIfNotPresent(interface{}) error
    HasSynced() bool
    Close()
}
```

Queue接口内嵌套了一个Store接口，Store定义在store.go中，代码如下：

```
type Store interface {
    Add(obj interface{}) error
    Update(obj interface{}) error
    Delete(obj interface{}) error
    List() []interface{}
    ListKeys() []string
    Get(obj interface{}) (item interface{}, exists bool, err error)
    GetByKey(key string) (item interface{}, exists bool, err error)
    Replace([]interface{}, string) error
    Resync() error
}
```

Store接口的方法基本可以望文生义，很直观，我们后面看一下在DeltaFIFO中是如何实现Store接口的。

2. DeltaFIFO结构体

```
type DeltaFIFO struct {
    lock sync.RWMutex
```

```
    cond sync.Cond
    items map[string]Deltas
    queue []string              // 这个queue里是没有重复元素的，和上面items的key保持一致
    populated bool
    initialPopulationCount int
    keyFunc KeyFunc             // 用于构造上面map用到的key
    knownObjects KeyListerGetter // 用来检索所有的key
    closed bool
    emitDeltaTypeReplaced bool
}
```

可以看到items属性是一个map，map的value是一个Deltas类型的，我们继续了解Deltas是怎么定义的：

```
type Delta struct {
    Type   DeltaType
    Object interface{}
}
type Deltas []Delta
```

Deltas是[]Delta类型的，Delta是一个结构体，其中的Type属性对应的DeltaType类型又是怎么定义的呢？

```
type DeltaType string

const (
    Added    DeltaType = "Added"
    Updated  DeltaType = "Updated"
    Deleted  DeltaType = "Deleted"
    Replaced DeltaType = "Replaced"
    Sync     DeltaType = "Sync"
)
```

原来DeltaType是一个字符串，对应的是用Added、Updated这种单词描述一个Delta的类型。将这些信息加在一起，我们可以尝试画出来DeltaFIFO的结构，大致如图5-3所示。

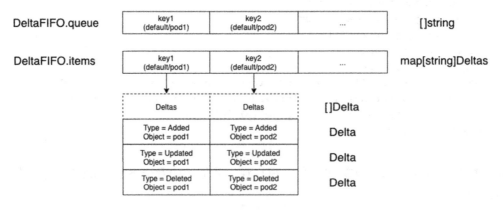

图 5-3 DeltaFIFO 的结构

首先 DeltaFIFO 结构体中有 queue 和 items 两个主要的属性，类型分别是 []string 和 map[string]Deltas，然后map[string]Deltas的key也就是default/pod1这种格式的字符串，这个后面

会看到，而value是类型为[]Delta的Deltas，这个Delta的属性也就是Type和Object，Type是前面提到的Added、Updated、Deleted这类字符串表示的DeltaType，Object就是这个Delta对应的对象，比如具体的某个Pod。

我们继续来看DeltaFIFO的New函数：

```
func NewDeltaFIFOWithOptions(opts DeltaFIFOOptions) *DeltaFIFO {
    if opts.KeyFunction == nil {
        opts.KeyFunction = MetaNamespaceKeyFunc
    }

    f := &DeltaFIFO{
        items:        map[string]Deltas{},
        queue:        []string{},
        keyFunc:      opts.KeyFunction,
        knownObjects: opts.KnownObjects,

        emitDeltaTypeReplaced: opts.EmitDeltaTypeReplaced,
    }
    f.cond.L = &f.lock
    return f
}
```

这里的逻辑比较简单。从这里可以看到一个MetaNamespaceKeyFunc函数，这个函数中可以看到前面提到的map[string]Deltas的key为什么是<namespace>/<name>这种格式的default/pod1。

5.3.2 queueActionLocked()方法的逻辑

在DeltaFIFO的实现中，Add()、Update()、Delete()等方法的画风基本都是这样的：

```
func (f *DeltaFIFO) Add(obj interface{}) error {
    f.lock.Lock()
    defer f.lock.Unlock()
    f.populated = true
    return f.queueActionLocked(Added, obj)
}
```

也就是说逻辑都落在了queueActionLocked()方法中，只是传递的参数不一样。所以我们关注的重点也就自然转移到了queueActionLocked()方法中了。

```
func (f *DeltaFIFO) queueActionLocked(actionType DeltaType, obj interface{}) error {
    id, err := f.KeyOf(obj) // 计算这个对象的key
    if err != nil {
        return KeyError{obj, err}
    }
    oldDeltas := f.items[id] // 从items map中获取当前的Deltas
    // 构造一个Delta，添加到Deltas中，也就是[]Delta中
    newDeltas := append(oldDeltas, Delta{actionType, obj})
    // 如果最近一个Delta是重复的，则保留后一个，目前版本只处理Deleted 的重复场景
    newDeltas = dedupDeltas(newDeltas)

    if len(newDeltas) > 0 {                        // 理论上newDeltas的长度一定大于0
```

```go
            if _, exists := f.items[id]; !exists {
                f.queue = append(f.queue, id)        // 如果id不存在，则在队列中添加
            }
            f.items[id] = newDeltas // 如果id已经存在，则只更新items map中对应这个key的Deltas
            f.cond.Broadcast()
        } else {
            // ...
            // 理论上这里的代码执行不到，省略掉不贴了
        }
        return nil
    }
```

现在再回头看Add()、Delete()、Update()、Get()等函数，就很简单了，它们只是将对应变化类型的obj添加到队列中而已。

5.3.3 Pop()方法和Replace()方法的逻辑

1. Pop()方法的实现

Pop()会按照元素的添加或更新顺序有序地返回一个元素（Deltas），在队列为空时会阻塞。另外，Pop过程会先从队列中删除一个元素后返回，所以如果处理失败了，则需要通过AddIfNotPresent()方法将这个元素加回到队列中。

Pop()的参数是type PopProcessFunc func(interface{}) error类型的process，在Pop()函数中，直接将队列中的第一个元素出队，然后丢给process处理，如果处理失败会重新入队，但是这个Deltas和对应的错误信息会被返回。

```go
func (f *DeltaFIFO) Pop(process PopProcessFunc) (interface{}, error) {
    f.lock.Lock()
    defer f.lock.Unlock()
    for {                          // 这个循环其实没有意义，和下面的!ok一起解决了一个不会发生的问题
        for len(f.queue) == 0 {                // 如果为空则进入这个循环
            if f.closed {                      // 如果队列关闭则直接返回
                return nil, ErrFIFOClosed
            }
            f.cond.Wait()                      // 等待
        }
        id := f.queue[0]                       // 在queue中放的是key
        f.queue = f.queue[1:]                  // queue中删除这个key
        depth := len(f.queue)
        if f.initialPopulationCount > 0 {      // 第一次调用Replace()插入的元素数量
            f.initialPopulationCount--
        }
        item, ok := f.items[id]        // 从items map[string]Deltas中获取一个Deltas
        if !ok {                       // 理论上不可能找不到，为此引入了上面的for嵌套
            klog.Errorf("Inconceivable! %q was in f.queue but not f.items; ignoring.", id)
            continue
```

```
            }
            delete(f.items, id)       // 在items map中也删除这个元素
            if depth > 10 {           // 当队列长度超过10并且处理一个元素的时间超过0.1秒时打印日
志。队列长度理论上不会变长，因为处理一个元素时是阻塞的，这时新的元素加不进来
                trace := utiltrace.New("DeltaFIFO Pop Process",
                    utiltrace.Field{Key: "ID", Value: id},
                    utiltrace.Field{Key: "Depth", Value: depth},
                    utiltrace.Field{Key: "Reason", Value: "slow event handlers
blocking the queue"})
                defer trace.LogIfLong(100 * time.Millisecond)
            }
            err := process(item)                    // 丢给PopProcessFunc处理
            if e, ok := err.(ErrRequeue); ok {      // 如果需要requeue则加回到队列中
                f.addIfNotPresent(id, item)
                err = e.Err
            }
            return item, err                        // 返回这个Deltas和错误信息
        }
    }
```

可以查看Pop()方法是如何被调用的，比如在当前包的controller.go中有这样一个方法：

```
func (c *controller) processLoop() {
    for {
        obj, err := c.config.Queue.Pop(PopProcessFunc(c.config.Process))
        if err != nil {
            if err == ErrFIFOClosed {
                return
            }
            if c.config.RetryOnError {
                // 其实Pop内部已经调用了AddIfNotPresent，这里重复调用一次并没有功能上的
帮助，不过可靠性提高了一点
                c.config.Queue.AddIfNotPresent(obj)
            }
        }
    }
}
```

看到这里，大家可能还有一个疑问，就是关于这里的process函数的实现逻辑是什么？我们在后面介绍Informer的时候会详细讲解sharedIndexInformer中如何实现process函数，所以暂时保留这个疑问。

2. Replace()方法的实现

Replace()方法简单地做了两件事：

1）给传入的对象列表添加一个Sync/Replace DeltaType的Delta。

2）执行一些与删除相关的程序逻辑。

这里的Replace()过程可以简单理解成传递一个新的[]Deltas过来，如果当前DeltaFIFO中已经有这些元素，则追加一个Sync/Replace动作，反之DeltaFIFO中多出来的Deltas可能与apiserver失

联导致实际被删除掉,但是删除事件并没有被监听(watch)到,所以直接追加一个类型为Deleted的Delta:

```go
func (f *DeltaFIFO) Replace(list []interface{}, _ string) error {
    f.lock.Lock()
    defer f.lock.Unlock()
    keys := make(sets.String, len(list))// 用来保存列表(list)中每个元素的key(键)
    action := Sync // 老代码兼容逻辑
    if f.emitDeltaTypeReplaced {
        action = Replaced
    }
    // 在每个item(元素)后面添加一个Sync/Replaced动作
    for _, item := range list {
        key, err := f.KeyOf(item)
        if err != nil {
            return KeyError{item, err}
        }
        keys.Insert(key)
        if err := f.queueActionLocked(action, item); err != nil {
            return fmt.Errorf("couldn't enqueue object: %v", err)
        }
    }

    if f.knownObjects == nil {
        queuedDeletions := 0
        for k, oldItem := range f.items { // 删除f.items中的旧元素
            if keys.Has(k) {
                continue
            }

            var deletedObj interface{}
            if n := oldItem.Newest(); n != nil { // 如果Deltas不为空则有返回值
                deletedObj = n.Object
            }
            queuedDeletions++
            // 标记删除,因为和apiserver失联引起的删除状态没有及时获取到,所以这里是DeletedFinalStateUnknown类型
            if err := f.queueActionLocked(Deleted, DeletedFinalStateUnknown{k, deletedObj}); err != nil {
                return err
            }
        }
        if !f.populated {
            f.populated = true
            f.initialPopulationCount = keys.Len() + queuedDeletions
        }
        return nil
    }
    // key就是如 "default/pod_1" 这种字符串
    knownKeys := f.knownObjects.ListKeys()
    queuedDeletions := 0
    for _, k := range knownKeys {
```

```
            if keys.Has(k) {
                continue
            }
            // 新列表中不存在的旧元素标记为将要删除
            deletedObj, exists, err := f.knownObjects.GetByKey(k)
            if err != nil {
                deletedObj = nil
                klog.Errorf("Unexpected error %v during lookup of key %v, placing
DeleteFinalStateUnknown marker without object", err, k)
            } else if !exists {
                deletedObj = nil
                klog.Infof("Key %v does not exist in known objects store, placing
DeleteFinalStateUnknown marker without object", k)
            }
            queuedDeletions++
            // 添加一个删除动作，因为与apiserver失联等场景会引起删除事件没有监听到，所以是
DeletedFinalStateUnknown类型
            if err := f.queueActionLocked(Deleted, DeletedFinalStateUnknown{k,
deletedObj}); err != nil {
                return err
            }
        }
    }

    if !f.populated {
        f.populated = true
        f.initialPopulationCount = keys.Len() + queuedDeletions
    }

    return nil
}
```

要完整理解Replace()的代码逻辑，还得看knownObjects的实现逻辑。如果继续去查看knownObjects属性的初始化逻辑，就可以看到其引用的是cache类型实现的Store，就是实现Indexer接口的一个实例，knownObjects通过cache类型的实例使用了和Indexer类似的机制，由内部ThreadSafeStore来实现检索队列所有元素的keys的能力。我们后面会详细介绍Indexer，到时大家回过头来看这里的代码逻辑应该就很清晰了。

5.4 Indexer 和 ThreadSafeStore

我们在前面讲过Indexer主要为对象提供根据一定条件进行检索的能力，典型的实现是通过namespace/name来构造key，通过ThreadSafeStore来存储对象。换言之，Indexer主要依赖于ThreadSafeStore实现，是client-go提供的一种缓存机制,通过检索本地缓存可以有效降低apiserver的压力。本节详细研究Indexer和对应的ThreadSafeStore的实现。

5.4.1 Indexer 接口和 cache 的实现

Indexer接口主要是在Store接口的基础上拓展了对象的检索功能，代码在k8s.io/client-go/

tools/cache包下。后文在不做特殊说明的情况下,提到的所有源文件都指的是这个包内的源文件。Indexer接口定义在index.go中:

```go
type Indexer interface {
    Store
    // 根据索引名和给定的对象返回符合条件的所有对象
    Index(indexName string, obj interface{}) ([]interface{}, error)
    // 根据索引名和索引值返回符合条件的所有对象的key
    IndexKeys(indexName, indexedValue string) ([]string, error)
    // 列出索引函数计算出来的所有索引值
    ListIndexFuncValues(indexName string) []string
    // 根据索引名和索引值返回符合条件的所有对象
    ByIndex(indexName, indexedValue string) ([]interface{}, error)
    // 获取所有的Indexers,对应map[string]IndexFunc类型
    GetIndexers() Indexers
    // 这个方法要在数据加入存储前调用,添加更多的索引方法,默认只通过 namespace检索
    AddIndexers(newIndexers Indexers) error
}
```

Indexer的默认实现是cache,cache定义在store.go中:

```go
type cache struct {
    cacheStorage ThreadSafeStore
    keyFunc KeyFunc
}
```

这里涉及两个类型:

- KeyFunc。
- ThreadSafeStore。

我们从Indexer的Add()方法和Update()方法的实现切入,看一下这两个类型的使用:

```go
func (c *cache) Add(obj interface{}) error {
    key, err := c.keyFunc(obj)
    if err != nil {
        return KeyError{obj, err}
    }
    c.cacheStorage.Add(key, obj)
    return nil
}
func (c *cache) Update(obj interface{}) error {
    key, err := c.keyFunc(obj)
    if err != nil {
        return KeyError{obj, err}
    }
    c.cacheStorage.Update(key, obj)
    return nil
}
```

可以看到这里的逻辑就是调用keyFunc()方法获取key,然后调用cacheStorage.Xxx()方法完成对应的增删改查过程。KeyFunc类型是这样定义的:

```go
type KeyFunc func(obj interface{}) (string, error)
```

也就是给一个对象返回一个字符串类型的key。KeyFunc的一个默认实现如下：

```go
func MetaNamespaceKeyFunc(obj interface{}) (string, error) {
    if key, ok := obj.(ExplicitKey); ok {
        return string(key), nil
    }
    meta, err := meta.Accessor(obj)
    if err != nil {
        return "", fmt.Errorf("object has no meta: %v", err)
    }
    if len(meta.GetNamespace()) > 0 {
        return meta.GetNamespace() + "/" + meta.GetName(), nil
    }
    return meta.GetName(), nil
}
```

可以看到一般情况下返回值是\<namespace>\<name>，如果namespace为空，则直接返回name。还有一个类型是ThreadSafeStore，我们放到下一小节来分析。

5.4.2 ThreadSafeStore 的实现

1. 接口与实现

ThreadSafeStore是Indexer的核心逻辑所在，Indexer的多数方法是直接调用内部cacheStorage属性的方法实现的，接口定义在thread_safe_store.go中：

```go
type ThreadSafeStore interface {
    Add(key string, obj interface{})
    Update(key string, obj interface{})
    Delete(key string)
    Get(key string) (item interface{}, exists bool)
    List() []interface{}
    ListKeys() []string
    Replace(map[string]interface{}, string)
    Index(indexName string, obj interface{}) ([]interface{}, error)
    IndexKeys(indexName, indexKey string) ([]string, error)
    ListIndexFuncValues(name string) []string
    ByIndex(indexName, indexKey string) ([]interface{}, error)
    GetIndexers() Indexers
    AddIndexers(newIndexers Indexers) error
    Resync() error // 过时废弃的方法
}
```

ThreadSafeStore对应的实现是threadSafeMap：

```go
type threadSafeMap struct {
    lock     sync.RWMutex
    items    map[string]interface{}
    indexers Indexers
```

```
    indices Indices
}
```

这里有一个Indexers和Indices，这两个类型分别是这样定义的：

```
type Index map[string]sets.String

type Indexers map[string]IndexFunc
type Indices map[string]Index
```

我们对照图5-4来理解这里的几个对象。

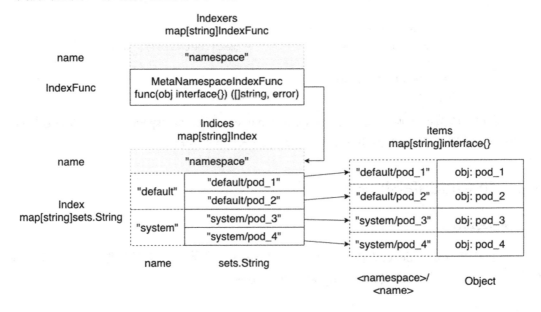

图 5-4　理解 IndexFunc、Indexers 和 Indices 几个对象

　　Indexers中保存的是Index函数map，一个典型的实现是字符串namespace作为key，IndexFunc类型的实现MetaNamespaceIndexFunc函数作为value，也就是我们希望通过 namespace来检索时，借助Indexers可以拿到对应的计算Index的函数，接着调用这个函数把对象传进去，就可以计算出这个对象对应的key，就是具体的namespace值，比如default、kube-system这种格式的字符串。然后在Indices中保存的也是一个map，key是上面计算出来的default这种格式的namespace值，value是一个set，而set表示的是这个default namespace下的一些具体pod的<namespace>/<name>这类格式字符串。最后拿着这个key，就可以在items中检索到对应的对象了。

2. Add()、Update()等方法的实现

我们继续看threadSafeMap如何实现添加元素，从Add()、Update()方法开始：

```
func (c *threadSafeMap) Add(key string, obj interface{}) {
    c.Update(key, obj)              // Add的实现就是直接调用Update
}
func (c *threadSafeMap) Update(key string, obj interface{}) {
    c.lock.Lock()
    defer c.lock.Unlock()
```

```go
    oldObject := c.items[key]              // c.items是map[string]interface{}类型
    c.items[key] = obj                     // 在items map中添加这个对象
    c.updateIndices(oldObject, obj, key)   // 下面分析这行代码
}
```

可以看到更复杂的逻辑在updateIndices()方法中，这个方法还是在thread_safe_store.go源文件中，我们继续查看：

```go
// 创建、更新、删除的入口都是这个方法，差异点在于create场景下的参数只传递 newObj，update场景
// 下需要传递oldObj和newObj，而delete场景只传递oldObj
func (c *threadSafeMap) updateIndices(oldObj interface{}, newObj interface{}, key string) {
    var oldIndexValues, indexValues []string
    var err error
    // 所有逻辑都在这个for循环中
    for name, indexFunc := range c.indexers {
        if oldObj != nil { // oldObj是否存在
            oldIndexValues, err = indexFunc(oldObj)
        } else { // 相当于置空操作
            oldIndexValues = oldIndexValues[:0]
        }
        // ...
        if newObj != nil { // newObj是否存在
            indexValues, err = indexFunc(newObj)
        } else { // 置空操作
            indexValues = indexValues[:0]
        }
        // ...
        // 拿到一个Index，对应类型map[string]sets.String
        index := c.indices[name]
        if index == nil {
            index = Index{} // 如果map不存在，则初始化一个
            c.indices[name] = index
        }
        // 处理oldIndexValues，也就是需要删除的索引值，这里保留了一个索引对应一个值的场景
        for _, value := range oldIndexValues {
            if len(indexValues) == 1 && value == indexValues[0] {
                continue
            }
            c.deleteKeyFromIndex(key, value, index)
        }
        // 处理indexValues，也就是需要添加的索引值，这里同样保留了一个索引对应一个值的场景
        for _, value := range indexValues {
            if len(oldIndexValues) == 1 && value == oldIndexValues[0] {
                continue
            }
            c.addKeyToIndex(key, value, index)
        }
    }
}
```

5.4.3 各种 Index 方法的实现

1. Index()方法

继续来看Index()方法的实现，其作用是给定一个obj和indexName，比如pod1和namespace，然后返回pod1所在namespace下的所有pod。同样在thread_safe_store.go源文件中：

```go
func (c *threadSafeMap) Index(indexName string, obj interface{}) ([]interface{}, error) {
    c.lock.RLock()
    defer c.lock.RUnlock()
    // 提取索引函数，比如通过namespace提取到 MetaNamespaceIndexFunc
    indexFunc := c.indexers[indexName]
    if indexFunc == nil {
        return nil, fmt.Errorf("Index with name %s does not exist", indexName)
    }
    // 对象丢进去拿到索引值，比如default
    indexedValues, err := indexFunc(obj)
    if err != nil {
        return nil, err
    }
    index := c.indices[indexName] // indexName如namespace，这里可以查到 Index

    var storeKeySet sets.String
    if len(indexedValues) == 1 { // 多数情况对应索引值为1的场景，比如用namespace时，值就是唯一的
        storeKeySet = index[indexedValues[0]]
    } else { // 对应索引值不为1的场景
        storeKeySet = sets.String{}
        for _, indexedValue := range indexedValues {
            for key := range index[indexedValue] {
                storeKeySet.Insert(key)
            }
        }
    }

    list := make([]interface{}, 0, storeKeySet.Len())
    // storeKey也就是default/pod_1这种字符串，通过其就可以到items map中提取需要的obj了
    for storeKey := range storeKeySet {
        list = append(list, c.items[storeKey])
    }
    return list, nil
}
```

2. ByIndex()方法

相较于Index()，ByIndex()函数要简单得多，直接传递 indexedValue，也就不需要通过 obj 去计算key了，例如indexName == namespace & indexValue == default就是直接检索 default下的资源对象。

```go
func (c *threadSafeMap) ByIndex(indexName, indexedValue string) ([]interface{}, error) {
```

```
    c.lock.RLock()
    defer c.lock.RUnlock()
    indexFunc := c.indexers[indexName]
    if indexFunc == nil {
        return nil, fmt.Errorf("Index with name %s does not exist", indexName)
    }
    index := c.indices[indexName]
    set := index[indexedValue]
    list := make([]interface{}, 0, set.Len())
    for key := range set {
        list = append(list, c.items[key])
    }
    return list, nil
}
```

3. IndexKeys()方法

和上面的方法返回obj列表的方式不同,这里只返回key列表,就是[]string{"default/pod_1"}这种格式的数据:

```
func (c *threadSafeMap) IndexKeys(indexName, indexedValue string) ([]string, error) {
    c.lock.RLock()
    defer c.lock.RUnlock()
    indexFunc := c.indexers[indexName]
    if indexFunc == nil {
        return nil, fmt.Errorf("Index with name %s does not exist", indexName)
    }
    index := c.indices[indexName]
    set := index[indexedValue]
    return set.List(), nil
}
```

5.5　ListerWatcher

ListerWatcher是Reflector的一个主要能力提供者,本节我们具体看一下ListerWatcher是如何实现List()和Watch()过程的。ListerWatcher的代码还是在k8s.io/client-go/tools/cache包中,后文在不明确提到其他包名的情况下,所有源文件都指的是这个包内的源文件。

5.5.1　ListWatch 对象的初始化

ListWatch对象和其创建过程都在listwatch.go中,我们先看一下ListWatch对象的定义:

```
type ListWatch struct {
    ListFunc   ListFunc
    WatchFunc  WatchFunc
```

```
        // 后面会介绍到chunk
        DisableChunking bool
}
```

可以看到这个结构体属性很简单，主要是ListFunc和WatchFunc。继续来看这个对象是怎么被初始化的：

```
// 这里Getter类型的c对应一个RESTClient
func NewListWatchFromClient(c Getter, resource string, namespace string,
fieldSelector fields.Selector) *ListWatch {
    optionsModifier := func(options *metav1.ListOptions) {
        options.FieldSelector = fieldSelector.String()// 序列化成字符串
    }
    // 调用下面这个NewFilteredListWatchFromClient()函数
    return NewFilteredListWatchFromClient(c, resource, namespace,
optionsModifier)
}
```

可以看到主要逻辑在NewFilteredListWatchFromClient()函数中，list和watch能力都是通过RESTClient提供的：

```
func NewFilteredListWatchFromClient(c Getter, resource string, namespace string,
optionsModifier func(options *metav1.ListOptions)) *ListWatch {
    // list某个namespace下的某个resource
    listFunc := func(options metav1.ListOptions) (runtime.Object, error) {
        optionsModifier(&options)
        return c.Get().// RESTClient.Get() -> *request.Request
            Namespace(namespace).
            Resource(resource).
            VersionedParams(&options, metav1.ParameterCodec).
            Do(context.TODO()).
            Get()
    }
    // 监听某个namespace（命名空间）下的资源
    watchFunc := func(options metav1.ListOptions) (watch.Interface, error) {
        options.Watch = true
        optionsModifier(&options)
        return c.Get().
            Namespace(namespace).
            Resource(resource).
            VersionedParams(&options, metav1.ParameterCodec).
            Watch(context.TODO())
    }
    return &ListWatch{ListFunc: listFunc, WatchFunc: watchFunc}
}
```

上面涉及一个Getter接口，我们继续看Getter的定义：

```
type Getter interface {
    Get() *restclient.Request
}
```

这里需要一个能够获得*restclient.Request的方式，我们实际使用时会用rest.Interface接口类型的实例，这是一个相对底层的工具，封装的是Kubernetes REST APIS相应的动作，可以在client-go的rest包内的client.go源文件中看到：

```go
type Interface interface {
    GetRateLimiter() flowcontrol.RateLimiter
    Verb(verb string) *Request
    Post() *Request
    Put() *Request
    Patch(pt types.PatchType) *Request
    Get() *Request
    Delete() *Request
    APIVersion() schema.GroupVersion
}
```

这个接口对应的实现也在client.go源文件中：

```go
type RESTClient struct {
    base *url.URL
    versionedAPIPath string
    content ClientContentConfig
    createBackoffMgr func() BackoffManager
    rateLimiter flowcontrol.RateLimiter
    warningHandler WarningHandler
    Client *http.Client
}
```

这里的RESTClient和平时Operator代码中常用的ClientSet的关系，可以通过这个简单的例子了解一下：我们在用ClientSet去Get一个指定名字的DaemonSet的时候，调用过程类似这样：

```go
r.AppsV1().DaemonSets("default").Get(ctx, "test-ds", getOpt)
```

这里的Get其实就是利用了RESTClient提供的能力，方法实现如下：

```go
func (c *daemonSets) Get(ctx context.Context, name string, options v1.GetOptions) (result *v1beta1.DaemonSet, err error) {
    result = &v1beta1.DaemonSet{}
    err = c.client.Get(). // 其实就是RESTClient.Get()，返回的是*rest.Request对象
        Namespace(c.ns).
        Resource("daemonsets").
        Name(name).
        VersionedParams(&options, scheme.ParameterCodec).
        Do(ctx).
        Into(result)
    return
}
```

5.5.2 ListerWatcher 接口

上面提到的ListWatch对象其实实现的是ListerWatcher接口，这个接口当然也在listwatch.go中定义：

```
type ListerWatcher interface {
    Lister
    Watcher
}
```

这里内嵌了两个接口，分别是Lister和Watcher：

```
type Lister interface {
    // List的返回值应该是一个list类型对象，也就是其中有Items字段，里面的 ResourceVersion
    // 可以用来监听（watch）
    List(options metav1.ListOptions) (runtime.Object, error)
}

type Watcher interface {
    // 从指定的资源版本开始watch
    Watch(options metav1.ListOptions) (watch.Interface, error)
}
```

ListWatch对象的List()和Watch()的实现就比较简短了：

```
func (lw *ListWatch) List(options metav1.ListOptions) (runtime.Object, error) {
    return lw.ListFunc(options)
}

func (lw *ListWatch) Watch(options metav1.ListOptions) (watch.Interface, error) {
    return lw.WatchFunc(options)
}
```

5.5.3　List-Watch 与 HTTP chunked

1. HTTP中的chunked

Kubernetes中主要通过List-Watch机制实现组件间的异步消息通信，List-Watch机制的实现原理值得深入分析。我们继续从HTTP层面来分析watch的实现机制，抓包试一下调用watch接口时数据包流向是怎样的。

Kubernetes中的监听（watch）长链接是通过HTTP的chunked机制实现的，在响应头中加一个Transfer-Encoding: chunked就可以实现分段响应。我们用Go语言来模拟一下这个过程，从而理解chunked是什么。这个Demo程序的服务器端代码如下：

```
func Server() {
    http.HandleFunc("/name", func(w http.ResponseWriter, r *http.Request) {
        flusher := w.(http.Flusher)
        for i := 0; i < 2; i++ {
            fmt.Fprintf(w, "Daniel Hu\n")
            flusher.Flush()
            <-time.Tick(1 * time.Second)
        }
    })
    log.Fatal(http.ListenAndServe(":8080", nil))
}
```

这里的逻辑是当客户端请求localhost:8080/name的时候，服务器端响应两行："Daniel Hu"。然后直接运行，再随便用一个工具进行访问，比如curl localhost:8080/name，接着抓包可以看到如图5-5所示的响应体。

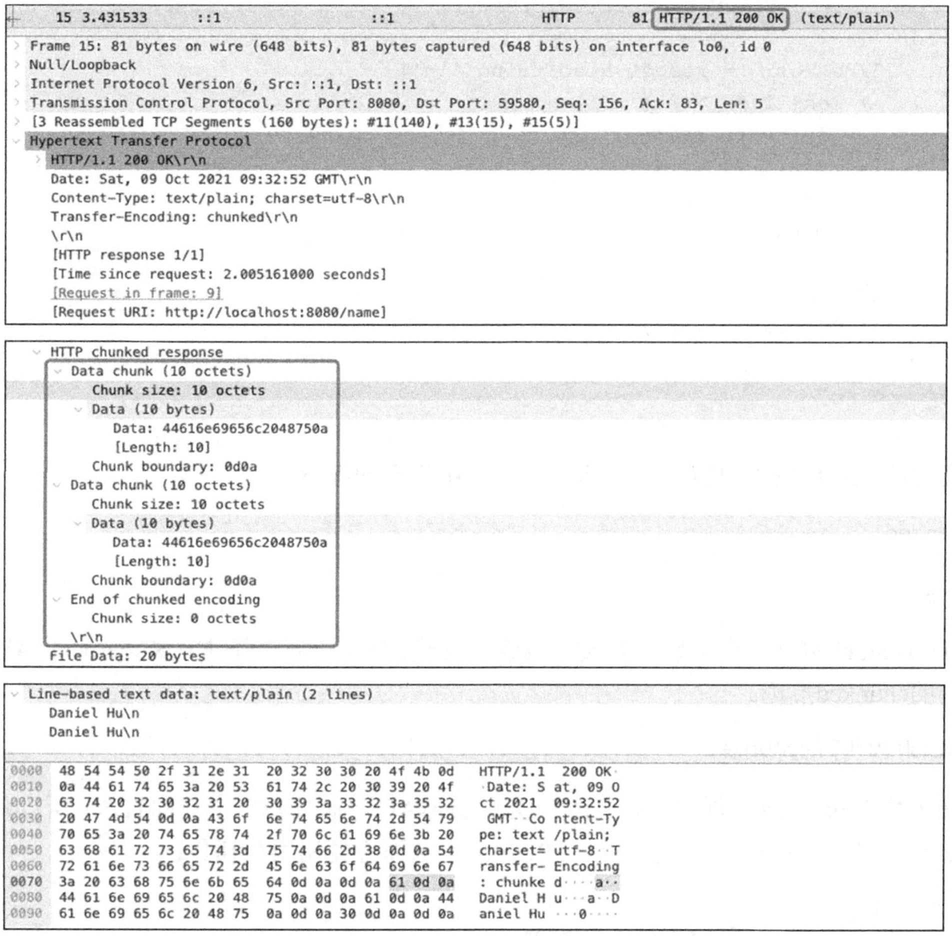

图 5-5 抓包看到的响应体

chunked类型的response由一个个chunk（块）组成，每个chunk的格式都是"Chunk size + Chunk data + Chunk boundary"，也就是块大小+数据+边界标识。chunk的结尾是一个大小为0的块，也就是"0\r\n"。串在一起整体格式类似这样：

```
[Chunk size][Chunk data][Chunk boundary][Chunk size][Chunk data][Chunk boundary][Chunk size=0][Chunk boundary]
```

在图5-5的例子中，服务器端响应的内容是两个相同的字符串"Daniel Hu\n"，客户端拿到的响应也就是"10Daniel Hu\r\n10Daniel Hu\r\n0\r\n"。

这种类型的数据怎么接收呢？可以这样：

```go
func Client() {
    resp, err := http.Get("http://127.0.0.1:8080/name")
    if err != nil {
        log.Fatal(err)
```

```go
    }
    defer resp.Body.Close()
    fmt.Println(resp.TransferEncoding)
    reader := bufio.NewReader(resp.Body)
    for {
        line, err := reader.ReadString('\n')
        if len(line) > 0 {
            fmt.Print(line)
        }
        if err == io.EOF {
            break
        }
        if err != nil {
            log.Fatal(err)
        }
    }
}
```

这段代码运行后输出内容如下（两个字符串中间会间隔1秒）：

```
[chunked]
Daniel Hu
Daniel Hu
```

HTTP的chunked类型响应数据方式大概就是这样，接下来看一下调用Kubernetes API时能否找到里面的chunked痕迹。

2. watch API中的chunked

现在多数Kubernetes集群都是以HTTPS方式暴露API，而且开启了双向TLS，所以我们需要先通过kubectl代理kube-apiserver提供HTTP的API，从而方便调用和抓包：

```
# kubectl proxy
Starting to serve on 127.0.0.1:8001
```

然后开始watch一个资源，比如这里选择coredns的configmap：

```
# curl localhost:8001/api/v1/watch/namespaces/kube-system/configmaps/coredns

{"type":"ADDED","object":{"kind":"ConfigMap","apiVersion":"v1","metadata":{"name":"coredns","namespace":"kube-system","uid":"f2eb4080-ce86-436a-9cfb-5bfdd3f59433","resourceVersion":"747388","creationTimestamp":"2022-02-06T02:49:56Z","managedFields":[{"manager":"kubeadm","operation":"Update","apiVersion":"v1","time":"2022-02-06T02:49:56Z","fieldsType":"FieldsV1","fieldsV1":{"f:data":{}}},{"manager":"kubectl-edit","operation":"Update","apiVersion":"v1","time":"2022-02-09T10:16:13Z","fieldsType":"FieldsV1","fieldsV1":{"f:data":{"f:Corefile":{}}}}]},"data":{"Corefile":".:53 {\n    errors\n    health {\n        lameduck 5s\n    }\n    ready\n    kubernetes cluster.local in-addr.arpa ip6.arpa {\n        pods insecure\n        fallthrough in-addr.arpa ip6.arpa\n        ttl 30\n    }\n    prometheus :9153\n    forward . /etc/resolv.conf {\n        max_concurrent 1000\n    }\n    cache 30\n    loop\n    reload\n    loadbalance\n}\n"}}}
```

这时可以马上拿到一个响应，然后我们通过kubectl命令去编辑一下这个configmap，可以看到监听端继续收到一条消息：

```
{"type":"MODIFIED","object":{"kind":"ConfigMap","apiVersion":"v1","metadata":{"name":"coredns","namespace":"kube-system","uid":"f2eb4080-ce86-436a-9cfb-5bfdd3f59433","resourceVersion":"747834","creationTimestamp":"2022-02-06T02:49:56Z","managedFields":[{"manager":"kubeadm","operation":"Update","apiVersion":"v1","time":"2022-02-06T02:49:56Z","fieldsType":"FieldsV1","fieldsV1":{"f:data":{}}},{"manager":"kubectl-edit","operation":"Update","apiVersion":"v1","time":"2022-02-09T10:16:13Z","fieldsType":"FieldsV1","fieldsV1":{"f:data":{"f:Corefile":{}}}}]},"data":{"Corefile":".:53 {\n    errors\n    health {\n        lameduck 5s\n    }\n    ready\n    kubernetes cluster.local in-addr.arpa ip6.arpa {\n        pods insecure\n        fallthrough in-addr.arpa ip6.arpa\n        ttl 30\n    }\n    prometheus :9153\n    forward . /etc/resolv.conf {\n        max_concurrent 1000\n    }\n    cache 31\n    loop\n    reload\n    loadbalance\n}\n"}}}
```

所以apiserver就是通过这样的方式将资源变更通知到各个watcher（监听器）的。这时如果我们去抓包，依旧可以看到这两个响应信息的具体数据包格式，第一个响应体如图5-6所示（截取了中间关键信息）。

```
// ……
0030    fe d7 b6 90 af fc 85 ac 48 54 54 50 2f 31 2e 31    ........HTTP/1.1
0040    20 32 30 30 20 4f 4b 0d 0a 41 75 64 69 74 2d 49     200 OK..Audit-I
// ……
0160    2d 62 66 38 38 66 37 66 39 66 33 66 61 0d 0a 54    -bf88f7f9f3fa..T
0170    72 61 6e 73 66 65 72 2d 45 6e 63 6f 64 69 6e 67    ransfer-Encoding
0180    3a 20 63 68 75 6e 6b 65 64 0d 0a 0d 0a 33 61 38    : chunked....3a8
0190    0d 0a 7b 22 74 79 70 65 22 3a 22 41 44 44 45 44    ..{"type":"ADDED
01a0    22 2c 22 6f 62 6a 65 63 74 22 3a 7b 22 6b 69 6e    ","object":{"kin
01b0    64 22 3a 22 43 6f 6e 66 69 67 4d 61 70 22 2c 22    d":"ConfigMap","
// ……
```

图5-6　第一个响应体

可以看到这里的HTTP头有一个Transfer-Encoding:chunked，下面的内容是 {"type":"ADDED…。继续看第二个包，这个包简单很多，少了HTTP头信息，只是简单地返回第二个chunk，如图5-7所示。

```
// ……
0030    fe d7 fb 0b af fc c0 4a 33 61 62 0d 0a 7b 22 74    .......J3ab..{"t
0040    79 70 65 22 3a 22 4d 4f 44 49 46 49 45 44 22 2c    ype":"MODIFIED",
0050    22 6f 62 6a 65 63 74 22 3a 7b 22 6b 69 6e 64 22    "object":{"kind"
0060    3a 22 43 6f 6e 66 69 67 4d 61 70 22 2c 22 61 70    :"ConfigMap","ap
// ……
```

图5-7　返回第二个chunk

这里可以看到0d 0a，也就是\r\n，至于前面的3ab，转成十进制就是939，对应这个chunk的长度，这个结果和前面自己写的HTTP服务"请求—响应"格式就一致了。

5.6 Reflector

我们前面提到过Reflector的任务就是从apiserver监听（watch）特定类型的资源，拿到变更通知后，将其丢到DeltaFIFO队列中，也介绍了ListerWatcher是如何从apiserver中列选-监听（List-Watch）资源的，本节我们继续来看Reflector的实现。

5.6.1 Reflector 的启动过程

Reflector定义在 k8s.io/client-go/tools/cache包下。下文在不做特殊说明的情况下，提到的所有源文件都指的是这个包内的源文件。代表Reflector的结构体属性比较多，我们在不知道其工作原理的情况下去逐个看这些属性意义不大，所以这里先不去具体看这个结构体的定义，而是直接找到Run()方法，从Reflector的启动切入，源码在reflector.go中：

```
func (r *Reflector) Run(stopCh <-chan struct{}) {
    klog.V(3).Infof("Starting reflector %s (%s) from %s", r.expectedTypeName,
r.resyncPeriod, r.name)
    wait.BackoffUntil(func() {
        if err := r.ListAndWatch(stopCh); err != nil {
            r.watchErrorHandler(r, err)
        }
    }, r.backoffManager, true, stopCh)
    klog.V(3).Infof("Stopping reflector %s (%s) from %s", r.expectedTypeName,
r.resyncPeriod, r.name)
}
```

这里有一些健壮性机制，用于处理apiserver短暂失联的场景，可以看到主要逻辑在Reflector.ListAndWatch()方法中。

5.6.2 核心方法：Reflector.ListAndWatch()

Reflector.ListAndWatch()方法有将近200行，是Reflector的核心逻辑之一。ListAndWatch()方法是先列出特定资源的所有对象，然后获取其资源版本，接着使用这个资源版本来开始监听流程。监听到新版本资源后，将其加入DeltaFIFO的动作是在watchHandler()方法中具体实现的。在此之前list（列选）到的最新元素会通过syncWith()方法添加一个Sync类型的DeltaType到DeltaFIFO中，所以list操作本身也会触发后面的调谐逻辑。具体代码如下：

```
func (r *Reflector) ListAndWatch(stopCh <-chan struct{}) error {
    // ...
    var resourceVersion string
    // 当r.lastSyncResourceVersion为""时这里是"0"，当使用r.lastSyncResourceVersion
失败时这里为""，区别是""会直接请求到etcd，获取一个新的版本，而"0"访问的是cache
    options := metav1.ListOptions{ResourceVersion: r.relistResourceVersion()}

    if err := func() error {
        // trace用于记录操作耗时，这里的逻辑是把超过10秒的步骤打印出来
```

```go
            initTrace := trace.New("Reflector ListAndWatch", trace.Field{Key: "name",
Value: r.name})
            defer initTrace.LogIfLong(10 * time.Second)
            var list runtime.Object
            var paginatedResult bool
            var err error
            listCh := make(chan struct{}, 1)
            panicCh := make(chan interface{}, 1)
            go func() {                          // 内嵌一个函数，这里会直接调用
                defer func() {
                    if r := recover(); r != nil {
                        panicCh <- r       // 收集这个goroutine的panic信息
                    }
                }()
                // 开始尝试收集list的chunks，我们前面介绍过chunk
                pager := pager.New(pager.SimplePageFunc(func(opts metav1.ListOptions)
(runtime.Object, error) {
                    return r.listerWatcher.List(opts)
                }))
                switch {
                case r.WatchListPageSize != 0:
                    pager.PageSize = r.WatchListPageSize
                case r.paginatedResult:
                case options.ResourceVersion != "" && options.ResourceVersion != "0":
                    pager.PageSize = 0
                }

                list, paginatedResult, err = pager.List(context.Background(),
options)
                if isExpiredError(err) || isTooLargeResourceVersionError(err) {
                    // 设置这个属性后，下一次list会从etcd中获取
                    r.setIsLastSyncResourceVersionUnavailable(true)
                    list, paginatedResult, err = pager.List(context.Background(),
metav1.ListOptions{ResourceVersion: r.relistResourceVersion()})
                }
                close(listCh)
            }()
            select {
            case <-stopCh:
                return nil
            case r := <-panicCh:
                panic(r)
            case <-listCh:
            }
            initTrace.Step("Objects listed", trace.Field{Key: "error", Value: err})
            if err != nil {
                klog.Warningf("%s: failed to list %v: %v", r.name, r.expectedTypeName,
err)
                return fmt.Errorf("failed to list %v: %v", r.expectedTypeName, err)
            }
            if options.ResourceVersion == "0" && paginatedResult {
                r.paginatedResult = true
```

```go
        }
        // list成功
        r.setIsLastSyncResourceVersionUnavailable(false) // list was successful
        listMetaInterface, err := meta.ListAccessor(list)
        if err != nil {
            return fmt.Errorf("unable to understand list result %#v: %v", list, err)
        }
        resourceVersion = listMetaInterface.GetResourceVersion()
        initTrace.Step("Resource version extracted")
        // 将list到的items添加到store中，这里的store也就是DeltaFIFO，也就是添加一个Sync
        // DeltaType，不过这里的resourceVersion值并没有实际用到
        items, err := meta.ExtractList(list)
        if err != nil {
            return fmt.Errorf("unable to understand list result %#v (%v)", list, err)
        }
        initTrace.Step("Objects extracted")
        if err := r.syncWith(items, resourceVersion); err != nil {
            return fmt.Errorf("unable to sync list result: %v", err)
        }
        initTrace.Step("SyncWith done")
        r.setLastSyncResourceVersion(resourceVersion)
        initTrace.Step("Resource version updated")
        return nil
    }(); err != nil {
        return err
    }

    resyncerrc := make(chan error, 1)
    cancelCh := make(chan struct{})
    defer close(cancelCh)
    go func() {
        resyncCh, cleanup := r.resyncChan()
        defer func() {
            cleanup() // Call the last one written into cleanup
        }()
        for {
            select {
            case <-resyncCh:
            case <-stopCh:
                return
            case <-cancelCh:
                return
            }
            if r.ShouldResync == nil || r.ShouldResync() {
                klog.V(4).Infof("%s: forcing resync", r.name)
                if err := r.store.Resync(); err != nil {
                    resyncerrc <- err
                    return
                }
            }
```

```go
                cleanup()
                resyncCh, cleanup = r.resyncChan()
            }
        }()

        for {
            select {
            case <-stopCh:
                return nil
            default:
            }
            // 超时时间是5~10分钟
            timeoutSeconds := int64(minWatchTimeout.Seconds() * (rand.Float64() + 1.0))
            options = metav1.ListOptions{
                ResourceVersion: resourceVersion,
                // 如果超时没有接收到任何Event，这时需要停止监听，避免一直阻塞
                TimeoutSeconds: &timeoutSeconds,
                // 用于降低apiserver压力，bookmark类型响应的对象主要只有RV信息
                AllowWatchBookmarks: true,
            }
            start := r.clock.Now()
            // 调用watch开始监听
            w, err := r.listerWatcher.Watch(options)
            if err != nil {
                // 这时直接re-list已经没有用了，apiserver暂时拒绝服务
                if utilnet.IsConnectionRefused(err) || apierrors.IsTooManyRequests(err) {
                    <-r.initConnBackoffManager.Backoff().C()
                    continue
                }
                return err
            }
            // 核心逻辑之一，后面会单独讲到watchHander()
            if err := r.watchHandler(start, w, &resourceVersion, resyncerrc, stopCh); err != nil {
                if err != errorStopRequested {
                    switch {
                    case isExpiredError(err):
                        // ...
                        <-r.initConnBackoffManager.Backoff().C()
                        continue
                    default:
                        // ...
                    }
                }
                return nil
            }
        }
    }
```

5.6.3 核心方法：Reflector.watchHandler()

前面分析ListAndWatch()方法时提到了会单独分析watchHandler()方法。在watchHandler()方法中完成了将监听到的Event（事件）根据其EventType（事件类型）分别调用DeltaFIFO的Add()/Update/Delete()等方法，完成对象追加到DeltaFIFO队列的过程。watchHandler()方法的调用在一个for循环中，所以一轮调用watchHandler()工作流程完成后函数退出，新一轮的调用会传递进来新的watch.Interface和resourceVersion等。我们来看watchHander()方法的实现，同样在reflector.go中：

```go
func (r *Reflector) watchHandler(start time.Time, w watch.Interface,
resourceVersion *string, errc chan error, stopCh <-chan struct{}) error {
    eventCount := 0
    // 当前函数返回时需要关闭watch.Interface，因为新一轮的调用会传递新的watch.Interface进来
    defer w.Stop()

loop:
    for {
        select {
        case <-stopCh:
            return errorStopRequested
        case err := <-errc:
            return err
        // 接收event（事件）
        case event, ok := <-w.ResultChan():
            if !ok {
                break loop
            }
            if event.Type == watch.Error { // 如果是Error（错误）
                return apierrors.FromObject(event.Object)
            }
            // 创建Reflector时会指定一个expectedType
            if r.expectedType != nil {
                // 类型不匹配
                if e, a := r.expectedType, reflect.TypeOf(event.Object); e != a {
                    utilruntime.HandleError(fmt.Errorf("%s: expected type %v, but watch event object had type %v", r.name, e, a))
                    continue
                }
            }
            // 没有对应Go语言结构体的对象可以通过这种方式来指定期望类型
            if r.expectedGVK != nil {
                if e, a := *r.expectedGVK, event.Object.GetObjectKind().GroupVersionKind(); e != a {
                    utilruntime.HandleError(fmt.Errorf("%s: expected gvk %v, but watch event object had gvk %v", r.name, e, a))
                    continue
                }
            }
            meta, err := meta.Accessor(event.Object)
            if err != nil {
```

```go
                    utilruntime.HandleError(fmt.Errorf("%s: unable to understand watch event %#v", r.name, event))
                    continue
                }
                // 新的ResourceVersion
                newResourceVersion := meta.GetResourceVersion()
                switch event.Type {
                // 调用DeltaFIFO的Add/Update/Delete等方法完成不同类型 Event的处理，我们在
// 5.3节详细介绍过DeltaFIFO对应的 Add/Update/Delete是如何实现的
                case watch.Added:
                    err := r.store.Add(event.Object)
                    if err != nil {
                        utilruntime.HandleError(fmt.Errorf("%s: unable to add watch event object (%#v) to store: %v", r.name, event.Object, err))
                    }
                case watch.Modified:
                    err := r.store.Update(event.Object)
                    if err != nil {
                        utilruntime.HandleError(fmt.Errorf("%s: unable to update watch event object (%#v) to store: %v", r.name, event.Object, err))
                    }
                case watch.Deleted:
                    err := r.store.Delete(event.Object)
                    if err != nil {
                        utilruntime.HandleError(fmt.Errorf("%s: unable to delete watch event object (%#v) from store: %v", r.name, event.Object, err))
                    }
                case watch.Bookmark:
                default:
                    utilruntime.HandleError(fmt.Errorf("%s: unable to understand watch event %#v", r.name, event))
                }
                // 更新resourceVersion
                *resourceVersion = newResourceVersion
                r.setLastSyncResourceVersion(newResourceVersion)
                if rvu, ok := r.store.(ResourceVersionUpdater); ok {
                    rvu.UpdateResourceVersion(newResourceVersion)
                }
                eventCount++
            }
        }
        watchDuration := r.clock.Since(start) // 耗时
        if watchDuration < 1*time.Second && eventCount == 0 { // 1秒就结束了，而且没有
// 收到事件，属于异常情况
            return fmt.Errorf("very short watch: %s: Unexpected watch close - watch lasted less than a second and no items received", r.name)
        }
        // ...
        return nil
    }
```

5.6.4 Reflector 的初始化

继续来看Reflector的初始化。NewReflector()的参数中有一个ListerWatcher类型的lw，还有一个expectedType和store，lw就是在ListerWatcher，expectedType指定期望关注的类型，而store是一个DeltaFIFO，加在一起大致可以预想到Reflector通过ListWatcher提供的能力去list-watch apiserver，然后完成将Event加到DeltaFIFO中。

还是在reflector.go中，我们查看相关的代码：

```go
func NewReflector(lw ListerWatcher, expectedType interface{}, store Store,
resyncPeriod time.Duration) *Reflector {
    // 直接调用下面的NewNamedReflector
    return NewNamedReflector(naming.GetNameFromCallsite(internalPackages...), lw,
expectedType, store, resyncPeriod)
}

func NewNamedReflector(name string, lw ListerWatcher, expectedType interface{},
store Store, resyncPeriod time.Duration) *Reflector {
    realClock := &clock.RealClock{}
    r := &Reflector{
        name:           name,
        listerWatcher:  lw,
        store:          store,
        // 重试机制，这里可以有效降低apiserver的负载，也就是重试间隔会越来越长
        backoffManager:         wait.NewExponentialBackoffManager
(800*time.Millisecond, 30*time.Second, 2*time.Minute, 2.0, 1.0, realClock),
        initConnBackoffManager: wait.NewExponentialBackoffManager
(800*time.Millisecond, 30*time.Second, 2*time.Minute, 2.0, 1.0, realClock),
        resyncPeriod:     resyncPeriod,
        clock:            realClock,
        watchErrorHandler:   WatchErrorHandler(DefaultWatchErrorHandler),
    }
    r.setExpectedType(expectedType)
    return r
}
```

5.6.5 小结

Reflector的职责很清晰，要做的事情是保持DeltaFIFO中的items持续更新，具体实现是通过ListerWatcher提供的list-watch（列选-监听）能力来列选指定类型的资源，这时会产生一系列Sync事件，然后通过列选到的ResourceVersion来开启监听过程，而监听到新的事件后，会和前面提到的Sync事件一样，都通过DeltaFIFO提供的方法构造相应的DeltaType添加到DeltaFIFO中。当然，前面提到的更新也并不是直接修改DeltaFIFO中已经存在的元素，而是添加一个新的DeltaType到队列中。另外，DeltaFIFO中添加新DeltaType时也会有一定的去重机制，我们在5.3节和5.5节中分别介绍过ListerWatcher和DeltaFIFO组件的工作逻辑，有了这个基础后再看Reflector的工作流就相对轻松很多了。

这里还有一个细节就是监听过程不是一劳永逸的，监听到新的事件后，会拿着对象的新

ResourceVersion重新开启一轮新的监听过程。当然，这里的watch调用也有超时机制，一系列的健壮性措施，所以我们脱离Reflector(Informer)直接使用list-watch还是很难直接写出一套健壮的代码逻辑。

5.7 Informer

在client-go源码分析的最后一节中，我们准备分析Informer。Informer这个词的出镜率很高，我们在很多文章中都可以看到Informer的身影，在源码中真的去找一个叫作Informer的对象，却又发现找不到一个单纯的Informer，但是有很多结构体或者接口中包含Informer这个词。

与Reflector、WorkQueue等组件不同，Informer相对来说更加模糊，让人初读源码时感觉迷惑。今天我们一起来揭开Informer的面纱，看一下它到底是什么。

我们在一开始提到过Informer从DeltaFIFO中Pop相应的对象，然后通过Indexer将对象和索引丢到本地cache中，再触发相应的事件处理函数（Resource Event Handlers）的运行。接下来我们通过源码来理解一下整个过程。

5.7.1 Informer 就是 Controller

1. Controller结构体与Controller接口

Informer通过一个Controller对象来定义，本身结构很简单，我们在k8s.io/client-go/tools/cache包中的controller.go源文件中可以看到Controller的定义：

```
type controller struct {
    config Config
    reflector *Reflector
    reflectorMutex sync.RWMutex
    clock clock.Clock
}
```

这里有我们熟悉的Reflector，可以猜到Informer启动时会去运行Reflector，从而通过Reflector实现list-watch apiserver，更新"事件"到DeltaFIFO中用于进一步处理。我们继续了解controller对应的Controller接口：

```
type Controller interface {
    Run(stopCh <-chan struct{})
    HasSynced() bool
    LastSyncResourceVersion() string
}
```

这里的核心方法很明显是Run(stopCh <-chan struct{})，Run负责两件事情：

1）构造Reflector利用ListerWatcher的能力将对象事件更新到DeltaFIFO。
2）从DeltaFIFO中Pop对象后调用ProcessFunc来处理。

2. Controller的初始化

同样，在controller.go文件中有如下代码：

```
func New(c *Config) Controller {
    ctlr := &controller{
        config: *c,
        clock:  &clock.RealClock{},
    }
    return ctlr
}
```

这里没有太多的逻辑，主要是传递了一个Config进来，可以猜到核心逻辑是Config从何而来以及后面如何使用。我们先向上跟踪Config从哪里来，New()的调用有几个地方，我们不去看newInformer()分支的代码，因为实际开发中主要是使用SharedIndexInformer，两个入口初始化Controller的逻辑类似，直接跟踪更实用的一个分支，查看func (s *sharedIndexInformer) Run(stopCh <-chan struct{})方法中如何调用New()，代码位于shared_informer.go中：

```
func (s *sharedIndexInformer) Run(stopCh <-chan struct{}) {
    // ...
    fifo := NewDeltaFIFOWithOptions(DeltaFIFOOptions{
        KnownObjects:          s.indexer,
        EmitDeltaTypeReplaced: true,
    })

    cfg := &Config{
        Queue:            fifo,
        ListerWatcher:    s.listerWatcher,
        ObjectType:       s.objectType,
        FullResyncPeriod: s.resyncCheckPeriod,
        RetryOnError:     false,
        ShouldResync:     s.processor.shouldResync,

        Process:          s.HandleDeltas,
        WatchErrorHandler: s.watchErrorHandler,
    }

    func() {
        s.startedLock.Lock()
        defer s.startedLock.Unlock()

        s.controller = New(cfg)
        s.controller.(*controller).clock = s.clock
        s.started = true
    }()
    // ...
    s.controller.Run(stopCh)
}
```

这里只保留了主要代码，后面会分析SharedIndexInformer，所以先不纠结SharedIndexInformer的细节，我们从这里可以看到SharedIndexInformer的Run()过程中会构造一个 Config，然后创建Controller，最后调用Controller的Run()方法。另外，这里也可以看到前面分析过的DeltaFIFO、ListerWatcher等，其中的Process:s.HandleDeltas这一行也比较重要，Process属性的类型是ProcessFunc，可以看到具体的ProcessFunc是HandleDeltas方法。

3. Controller的启动

上面提到Controller的初始化本身没有太多的逻辑，主要是构造了一个Config对象传递进来，所以Controller启动时肯定会有这个Config的使用逻辑。我们回到controller.go文件具体查看：

```go
func (c *controller) Run(stopCh <-chan struct{}) {
    // ...
    r := NewReflector(// 利用Config中的配置构造Reflector
        c.config.ListerWatcher,
        c.config.ObjectType,
        c.config.Queue,
        c.config.FullResyncPeriod,
    )
    r.ShouldResync = c.config.ShouldResync
    r.WatchListPageSize = c.config.WatchListPageSize
    r.clock = c.clock
    if c.config.WatchErrorHandler != nil {
        r.watchErrorHandler = c.config.WatchErrorHandler
    }

    c.reflectorMutex.Lock()
    c.reflector = r
    c.reflectorMutex.Unlock()

    var wg wait.Group
    // 启动Reflector
    wg.StartWithChannel(stopCh, r.Run)
    // 执行Controller的processLoop
    wait.Until(c.processLoop, time.Second, stopCh)
    wg.Wait()
}
```

这里的代码逻辑很简单，构造Reflector后运行起来，然后执行c.processLoop，显然Controller的业务逻辑隐藏在processLoop方法中。我们继续来看processLoop的代码逻辑。

4. processLoop

```go
func (c *controller) processLoop() {
    for {
        obj, err := c.config.Queue.Pop(PopProcessFunc(c.config.Process))
        if err != nil {
            if err == ErrFIFOClosed {
                return
            }
            if c.config.RetryOnError {
                // This is the safe way to re-enqueue.
                c.config.Queue.AddIfNotPresent(obj)
            }
        }
    }
}
```

这里的代码逻辑是从DeltaFIFO中Pop出一个对象丢给PopProcessFunc处理，如果失败了就re-enqueue到DeltaFIFO中。我们前面提到过这里的PopProcessFunc由HandleDeltas()方法来实现，所以这里的主要逻辑就转到了HandleDeltas()是如何实现的。

5. HandleDeltas()

如果大家记不清DeltaFIFO的存储结构，可以回到前面相关章节看一下DeltaFIFO的结构图，然后回到这里查看源码。代码位于shared_informer.go文件中：

```go
func (s *sharedIndexInformer) HandleDeltas(obj interface{}) error {
    s.blockDeltas.Lock()
    defer s.blockDeltas.Unlock()

    if deltas, ok := obj.(Deltas); ok {
        return processDeltas(s, s.indexer, s.transform, deltas)
    }
    return errors.New("object given as Process argument is not Deltas")
}
```

代码逻辑都落在processDeltas()函数的调用上，我们继续看下面的代码：

```go
func processDeltas(
    handler ResourceEventHandler,
    clientState Store,
    transformer TransformFunc,
    deltas Deltas,
) error {
    // 对于每个Deltas来说，其中保存了很多Delta，也就是对应不同类型的多个对象，这里的遍历会从旧往新走
    for _, d := range deltas {
        obj := d.Object
        // ...
        switch d.Type { // 除了 Deleted 外的所有情况
        case Sync, Replaced, Added, Updated:
            if old, exists, err := clientState.Get(obj); err == nil && exists {
                // 通过indexer从cache中查询当前Object，如果存在则更新indexer中的对象
                if err := clientState.Update(obj); err != nil {
                    return err
                }
                // 调用ResourceEventHandler的OnUpdate()
                handler.OnUpdate(old, obj)
            } else {
                // 将对象添加到indexer中
                if err := clientState.Add(obj); err != nil {
                    return err
                }
                // 调用ResourceEventHandler的OnAdd()
                handler.OnAdd(obj)
            }
        case Deleted:
            // 如果是删除操作，则从indexer中删除这个对象
            if err := clientState.Delete(obj); err != nil {
                return err
            }
            // 调用ResourceEventHandler的OnDelete()
```

```
            handler.OnDelete(obj)
        }
    }
    return nil
}
```

这里的代码逻辑主要是遍历一个Deltas中的所有Delta，然后根据Delta的类型来决定如何操作Indexer，也就是更新本地cache，同时分发相应的通知。

5.7.2 SharedIndexInformer 对象

1. SharedIndexInformer是什么

在Operator开发中，如果不使用controller-runtime库，也就是不通过Kubebuilder等工具来生成脚手架，经常会用到SharedInformerFactory，比如典型的sample-controller中的main()函数：

```
func main() {
    // ...
    kubeClient, err := kubernetes.NewForConfig(cfg)
    // ...
    exampleClient, err := clientset.NewForConfig(cfg)
    // ...
    kubeInformerFactory := kubeinformers.NewSharedInformerFactory(kubeClient,
time.Second*30)
    exampleInformerFactory := informers.NewSharedInformerFactory(exampleClient,
time.Second*30)

    controller := NewController(kubeClient, exampleClient,
        kubeInformerFactory.Apps().V1().Deployments(),
        exampleInformerFactory.Samplecontroller().V1alpha1().Foos())

    kubeInformerFactory.Start(stopCh)
    exampleInformerFactory.Start(stopCh)

    if err = controller.Run(2, stopCh); err != nil {
        klog.Fatalf("Error running controller: %s", err.Error())
    }
}
```

这里可以看到我们依赖于kubeInformerFactory.Apps().V1().Deployments()提供一个 Informer，其中的Deployments()方法返回的是DeploymentInformer类型，DeploymentInformer又是什么呢？我们继续往下看，在client-go的informers/apps/v1包的deployment.go文件中有相关定义：

```
type DeploymentInformer interface {
    Informer() cache.SharedIndexInformer
    Lister() v1.DeploymentLister
}
```

可以看到所谓的DeploymentInformer是由Informer和Lister组成的，也就是说平时编码时用到的Informer本质就是一个SharedIndexInformer。

2. SharedIndexInformer接口的定义

回到shared_informer.go文件中,可以看到SharedIndexInformer接口的定义:

```go
type SharedIndexInformer interface {
    SharedInformer
    AddIndexers(indexers Indexers) error
    GetIndexer() Indexer
}
```

这里的Indexer就很熟悉了,SharedInformer又是什么呢?我们继续往下看:

```go
type SharedInformer interface {
    // 可以添加自定义的ResourceEventHandler
    AddEventHandler(handler ResourceEventHandler)
    // 附带resync间隔配置,设置为0表示不关心resync
    AddEventHandlerWithResyncPeriod(handler ResourceEventHandler, resyncPeriod time.Duration)
    // 这里的Store指的是Indexer
    GetStore() Store
    GetController() Controller                  // 过时了,没有用
    Run(stopCh <-chan struct{})                 // 通过Run来启动
    // 这里和resync逻辑没有关系,表示Indexer至少更新过一次全量的对象
    HasSynced() bool
    LastSyncResourceVersion() string            // 最后一次拿到的RV
    // 用于每次ListAndWatch连接断开时回调,主要是日志记录的作用
    SetWatchErrorHandler(handler WatchErrorHandler) error
    // 用于在对象存储前执行一些操作
    SetTransform(handler TransformFunc) error
}
```

3. sharedIndexInformer结构体的定义

继续来看SharedIndexInformer接口的实现sharedIndexerInformer是如何定义的,同样在shared_informer.go文件中查看代码:

```go
type sharedIndexInformer struct {
    indexer    Indexer
    controller Controller
    processor            *sharedProcessor
    cacheMutationDetector MutationDetector
    listerWatcher ListerWatcher
    // 表示当前Informer期望关注的类型,主要是GVK信息
    objectType runtime.Object

    // reflector的resync计时器计时间隔,通知所有的listener执行resync
    resyncCheckPeriod time.Duration
    defaultEventHandlerResyncPeriod time.Duration
    clock clock.Clock
    started, stopped bool
    startedLock sync.Mutex
    blockDeltas sync.Mutex
    watchErrorHandler WatchErrorHandler
    transform TransformFunc
}
```

这里的Indexer、Controller、ListerWatcher等都是熟悉的组件，sharedProcessor在前面已经遇到过，这也是一个需要关注的重点逻辑，5.7.3节专门来分析sharedProcessor的实现逻辑。

4. sharedIndexInformer的启动

继续来看sharedIndexInformer的Run()方法，其代码在shared_informer.go文件中，这里除了将在5.7.3节介绍的sharedProcessor外，几乎已经没有陌生的内容了：

```go
func (s *sharedIndexInformer) Run(stopCh <-chan struct{}) {
    defer utilruntime.HandleCrash()
    // ...
    // DeltaFIFO我们很熟悉了
    fifo := NewDeltaFIFOWithOptions(DeltaFIFOOptions{
        KnownObjects:        s.indexer,
        EmitDeltaTypeReplaced: true,
    })
    // Config 的逻辑也在上面遇到过了
    cfg := &Config{
        Queue:             fifo,
        ListerWatcher:     s.listerWatcher,
        ObjectType:        s.objectType,
        FullResyncPeriod:  s.resyncCheckPeriod,
        RetryOnError:      false,
        ShouldResync:      s.processor.shouldResync,

        Process:           s.HandleDeltas,
        WatchErrorHandler: s.watchErrorHandler,
    }

    func() {
        s.startedLock.Lock()
        defer s.startedLock.Unlock()

        s.controller = New(cfg)      // 前文分析过这个New()函数逻辑了
        s.controller.(*controller).clock = s.clock
        s.started = true
    }()

    processorStopCh := make(chan struct{})
    var wg wait.Group
    defer wg.Wait()                          // 等待Processor结束
    defer close(processorStopCh)             // 告诉Processor可以结束了
    wg.StartWithChannel(processorStopCh, s.cacheMutationDetector.Run)
    // processor的run方法
    wg.StartWithChannel(processorStopCh, s.processor.run)

    defer func() {
        s.startedLock.Lock()
        defer s.startedLock.Unlock()
        s.stopped = true         // 从而拒绝新的listener
    }()
    s.controller.Run(stopCh)     // controller的Run()
}
```

5.7.3　sharedProcessor 对象

sharedProcessor中维护了processorListener集合，然后分发通知对象到listeners，先研究结构定义，其代码在shared_informer.go中：

```go
type sharedProcessor struct {
    listenersStarted bool
    listenersLock    sync.RWMutex
    listeners        []*processorListener
    syncingListeners []*processorListener
    clock            clock.Clock
    wg               wait.Group
}
```

这里可以看到一个processorListener类型，下面看一下这个类型是怎么定义的：

```go
type processorListener struct {
    nextCh chan interface{}
    addCh  chan interface{}
    // 核心属性
    handler ResourceEventHandler
    pendingNotifications buffer.RingGrowing
    requestedResyncPeriod time.Duration
    resyncPeriod time.Duration
    nextResync time.Time
    resyncLock sync.Mutex
}
```

可以看到processorListener中有一个ResourceEventHandler，这是我们认识的组件。processorListener有三个主要方法：

- run()。
- add(notification interface{})。
- pop()。

我们逐一来看这三个方法的实现：

1. run()

```go
func (p *processorListener) run() {
    stopCh := make(chan struct{})
    wait.Until(func() {
        for next := range p.nextCh {
            switch notification := next.(type) {
            case updateNotification:
                p.handler.OnUpdate(notification.oldObj, notification.newObj)
            case addNotification:
                p.handler.OnAdd(notification.newObj)
            case deleteNotification:
                p.handler.OnDelete(notification.oldObj)
            default:
```

```
                        utilruntime.HandleError(fmt.Errorf("unrecognized notification: %T",
next))
                }
        }
        close(stopCh)
    }, 1*time.Second, stopCh)
}
```

这里的逻辑很清晰，从nextCh中拿通知，然后根据其类型去调用ResourceEventHandler相应的OnAdd()/OnUpdate()/OnDelete()方法。

2. add()和pop()

```
func (p *processorListener) add(notification interface{}) {
    // 将通知放到addCh中，所以下面的pop()方法中先执行到的case是第二个
    p.addCh <- notification
}

func (p *processorListener) pop() {
    defer utilruntime.HandleCrash()
    defer close(p.nextCh)              // 通知run()结束运行

    var nextCh chan<- interface{}
    var notification interface{}
    for {
        select {
        // 下面将获取到的通知添加到nextCh中，供run()方法中消费
        case nextCh <- notification:
            // 分发通知
            var ok bool
            // 从pendingNotifications中消费通知，生产者在下面的case中
            notification, ok = p.pendingNotifications.ReadOne()
            if !ok {           // 没有内容可以pop
                nextCh = nil
            }
        // 逻辑从这里开始，从addCh中提取通知
        case notificationToAdd, ok := <-p.addCh:
            if !ok {
                return
            }
            if notification == nil {
                notification = notificationToAdd
                nextCh = p.nextCh
            } else {            // 新添加的通知丢到pendingNotifications中
                p.pendingNotifications.WriteOne(notificationToAdd)
            }
        }
    }
}
```

可以看到processorListener提供了一定的缓冲机制来接收notification，然后去消费这些notification调用ResourceEventHandler相关方法。接下来我们继续看sharedProcessor的几种主要方法：

- addListener()。
- distribute()。
- run()。

addListener()方法会直接调用前面讲过的listener的run()和pop()方法，我们来看具体的代码：

```
func (p *sharedProcessor) addListener(listener *processorListener) {
    p.listenersLock.Lock()
    defer p.listenersLock.Unlock()

    p.addListenerLocked(listener)
    if p.listenersStarted {
        p.wg.Start(listener.run)
        p.wg.Start(listener.pop)
    }
}
```

而distribute()方法的逻辑就是调用sharedProcessor内部维护的所有listener的add()方法：

```
func (p *sharedProcessor) distribute(obj interface{}, sync bool) {
    p.listenersLock.RLock()
    defer p.listenersLock.RUnlock()

    if sync {
        for _, listener := range p.syncingListeners {
            listener.add(obj)
        }
    } else {
        for _, listener := range p.listeners {
            listener.add(obj)
        }
    }
}
```

最后查看run()方法的逻辑，这个方法和前面的addListener()方法类似，也就是调用listener的run()和pop()方法：

```
func (p *sharedProcessor) run(stopCh <-chan struct{}) {
    func() {
        p.listenersLock.RLock()
        defer p.listenersLock.RUnlock()
        for _, listener := range p.listeners {
            p.wg.Start(listener.run)
            p.wg.Start(listener.pop)
        }
        p.listenersStarted = true
    }()
    <-stopCh
    p.listenersLock.RLock()
    defer p.listenersLock.RUnlock()
    for _, listener := range p.listeners {
        // 通知pop()结束运行；pop()方法会通知run()结束运行
        close(listener.addCh)
    }
```

```
        p.wg.Wait() // 等待所有的pop()和run()方法运行结果
}
```

至此，我们基本就知道sharedProcessor的能力了。另外，SharedIndexInformer的逻辑也就基本讲完了，再往上层看这套代码逻辑，就剩下一个 SharedInformerFactory 了，5.7.4节继续来了解它。

5.7.4 关于SharedInformerFactory

SharedInformerFactory是在开发Operator的过程中经常会接触到的一个比较高层的抽象对象，接下来开始详细分析这个对象的源码。SharedInformerFactory定义在k8s.io/client-go/tools/cache包下。下文在不做特殊说明的情况下，提到的所有源文件都指的是这个包内的源文件。

1. SharedInformerFactory的定义

SharedInformerFactory接口定义在factory.go文件中：

```
type SharedInformerFactory interface {
    internalinterfaces.SharedInformerFactory
    ForResource(resource schema.GroupVersionResource) (GenericInformer, error)
    WaitForCacheSync(stopCh <-chan struct{}) map[reflect.Type]bool

    Admissionregistration() admissionregistration.Interface
    Internal() apiserverinternal.Interface
    Apps() apps.Interface
    Autoscaling() autoscaling.Interface
    Batch() batch.Interface
    Certificates() certificates.Interface
    Coordination() coordination.Interface
    Core() core.Interface
    Discovery() discovery.Interface
    Events() events.Interface
    Extensions() extensions.Interface
    Flowcontrol() flowcontrol.Interface
    Networking() networking.Interface
    Node() node.Interface
    Policy() policy.Interface
    Rbac() rbac.Interface
    Scheduling() scheduling.Interface
    Storage() storage.Interface
}
```

这里可以看到一个internalinterfaces.SharedInformerFactory接口，我们看一下这个接口的定义，代码在internalinterfaces/factory_interfaces.go中：

```
type SharedInformerFactory interface {
    Start(stopCh <-chan struct{})
    InformerFor(obj runtime.Object, newFunc NewInformerFunc) cache.SharedIndexInformer
}
```

这里可以看到熟悉的SharedIndexInformer。

然后了解ForResource(resource schema.GroupVersionResource) (GenericInformer, error)这行代码的逻辑，这里接收一个GVR，返回了一个GenericInformer。什么是GenericInformer呢？我们在generic.go中可以看到相应的定义：

```
type GenericInformer interface {
    Informer() cache.SharedIndexInformer
    Lister() cache.GenericLister
}
```

接着看SharedInformerFactory接口剩下的一大堆相似的方法，我们以Apps() apps.Interface为例，这个Interface定义在apps/interface.go中：

```
type Interface interface {
    V1() v1.Interface
    V1beta1() v1beta1.Interface
    V1beta2() v1beta2.Interface
}
```

这时大家的关注点肯定是这里的v1.Interface类型是怎么定义的，继续到apps/v1/interface.go文件中查看：

```
type Interface interface {
    ControllerRevisions() ControllerRevisionInformer
    DaemonSets() DaemonSetInformer
    Deployments() DeploymentInformer
    ReplicaSets() ReplicaSetInformer
    StatefulSets() StatefulSetInformer
}
```

到这里已经有看着很眼熟的Deployments() DeploymentInformer之类的代码了，前面看过DeploymentInformer的定义：

```
type DeploymentInformer interface {
    Informer() cache.SharedIndexInformer
    Lister() v1.DeploymentLister
}
```

现在也就不难理解SharedInformerFactory的作用了，它提供了所有API group-version的资源对应的SharedIndexInformer，也就不难理解前面引用的sample-controller中的这行代码：

```
kubeInformerFactory.Apps().V1().Deployments()
```

通过其可以拿到一个Deployment资源对应的SharedIndexInformer。

2. SharedInformerFactory的初始化

继续来看SharedInformerFactory的New()逻辑，其代码在factory.go中：

```
func NewSharedInformerFactory(client kubernetes.Interface, defaultResync time.Duration) SharedInformerFactory {
    return NewSharedInformerFactoryWithOptions(client, defaultResync)
}
```

可以看到参数非常简单，主要是需要一个ClientSet，毕竟ListerWatcher的能力本质还是client提供的。继续来看这里调用的NewSharedInformerFactoryWithOptions()函数：

```
func NewSharedInformerFactoryWithOptions(client kubernetes.Interface,
defaultResync time.Duration, options ...SharedInformerOption) SharedInformerFactory {
    factory := &sharedInformerFactory{
        client:          client,
        namespace:       v1.NamespaceAll, // 空字符串""
        defaultResync:   defaultResync,
        // 可以存放不同类型的SharedIndexInformer
        informers:        make(map[reflect.Type]cache.SharedIndexInformer),
        startedInformers: make(map[reflect.Type]bool),
        customResync:     make(map[reflect.Type]time.Duration),
    }

    for _, opt := range options {
        factory = opt(factory)
    }

    return factory
}
```

3. SharedInformerFactory的启动过程

最后查看SharedInformerFactory是如何启动的，Start()方法同样位于factory.go源文件中：

```
func (f *sharedInformerFactory) Start(stopCh <-chan struct{}) {
    f.lock.Lock()
    defer f.lock.Unlock()

    for informerType, informer := range f.informers {
        // 同类型只会调用一次，Run()的代码逻辑前面介绍过了
        if !f.startedInformers[informerType] {
            go informer.Run(stopCh)
            f.startedInformers[informerType] = true
        }
    }
}
```

5.7.5 小结

我们从一个基础Informer-Controller开始介绍，先分析了Controller的能力，其通过构造Reflector并启动从而能够获取指定类型资源的"更新"事件，然后通过事件构造Delta放到DeltaFIFO中，进而在processLoop中从DeltaFIFO里pop Deltas来处理，一方面将对象通过Indexer同步到本地cache（也就是一个ThreadSafeStore），另一方面调用ProcessFunc来处理这些Delta。

然后SharedIndexInformer提供了构造Controller的能力，通过HandleDeltas()方法实现上面提到的ProcessFunc，同时还引入了sharedProcessor在HandleDeltas()中用于事件通知的处理。sharedProcessor分发事件通知时，接收方是内部继续抽象出来的processorListener，在processorListener中完成了ResourceEventHandler回调函数的具体调用。

最后SharedInformerFactory又进一步封装了提供所有API资源对应的SharedIndexInformer的能力。也就是说一个SharedIndexInformer可以处理一种类型的资源，比如Deployment或者Pod等，而通过SharedInformerFactory可以轻松构造任意已知类型的SharedIndexInformer。另外，这里用到了ClientSet提供的访问所有API资源的能力，通过它也就能够完整实现整套Informer程序逻辑了。

5.8 本章小结

本章我们详细分析了client-go中与Operator开发相关的各种组件的原理与源码，这些知识虽然掌握起来有一定的难度，但是对我们开发一个复杂的Operator程序或者对平时定位Operator程序中遇到的各种问题是很有帮助的。我们希望能够通过对client-go源码的学习，进而对Operator的原理细节知根知底，从而实现在工作中开发、调试Operator项目游刃有余。

第 6 章
项目核心依赖包分析

我们在开发Operator项目时，除了用到client-go外，还会依赖很多k8s.io下的项目。本章将说明几个主要依赖项目分别都用于解决什么问题。

6.1 API 项目

API项目的GitHub地址是https://github.com/kubernetes/api。从项目名称上我们大概就可以猜到这个项目中存放的是Kubernetes的API定义。

API的定义单独放到一个项目中是为了解决循环依赖问题。需要使用到API定义的项目主要是k8s.io/client-go、k8s.io/apimachinery和k8s.io/apiserver，当然我们自己开发的Operator项目中也不可避免地需要引用一些API对象。

从API项目的目录结构可以很容易找到熟悉的内容：

```
...
apps/
autoscaling/
batch/
core/
discovery/
events/
extensions/
networking/
storage/
...
```

这里只列出了一部分目录，可以看到这些都是API Group的名字。我们以apps为线索继续查看其中的内容：

```
v1/
v1beta1/
v1beta2/
```

没错，Group里面不出所料是按照Version来划分的目录。那么v1目录中应该就是apps/v1对应的所有Kind的定义了，我们找一下有没有熟悉的内容，如图6-1所示。

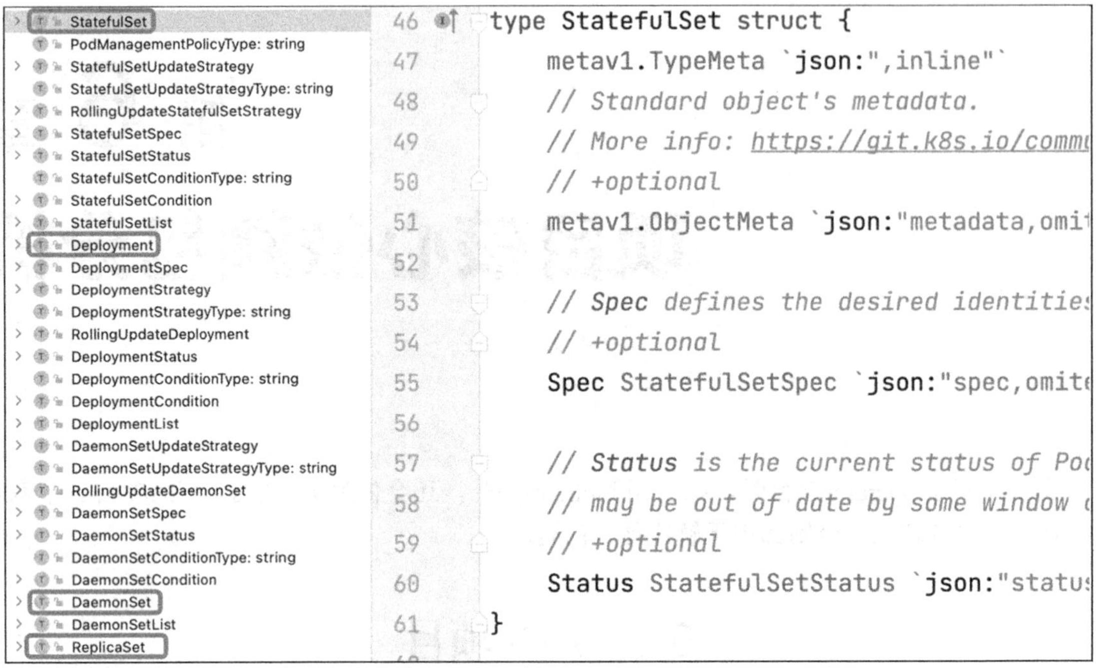

图 6-1 一些常见的类型定义

如图6-1所示，这里有很多我们平时经常见的类型定义，比如StatefulSet、Deployment、DaemonSet、ReplicaSet等。所以以后我们在Operator开发的过程中想要操作某个API，但是又不清楚这个API资源定义去哪里引用，记住到k8s.io/api项目中查看。

另外，k8s.io/api项目和k8s.io/client-go同样是从k8s.io/kubernetes项目的staging下同步过来的，也就是说给k8s.io/api项目贡献代码同样需要提交到Kubernetes主库。

6.2 apimachinery 项目

apimachinery项目的GitHub地址是https://github.com/kubernetes/apimachinery。machinery是"机械、组织、体制、系统"的意思，从项目名称上大致可以猜到这个项目实现的是各种和API相关操作的封装。

apimachinery项目的作用是为了解耦用到Kubernetes API的服务端和客户端，实现了很多公共类型依赖，主要包含Scheme、类型转换、编码解码等逻辑。依赖apimachinery的项目主要是k8s.io/kubernetes、k8s.io/client-go和k8s.io/apiserver等，当然同样我们自己开发的Operator项目中也不可避免地需要用到这个项目的一些代码。

k8s.io/apimachinery 项目同样也是从 k8s.io/kubernetes 项目的 staging 下同步过来的，给 k8s.io/apimachinery项目贡献代码同样需要提交到Kubernetes主库。

6.3 controller-runtime 项目

controller-runtime包对应的项目地址在https://github.com/kubernetes-sigs/controller-runtime。与前面介绍的k8s.io/api和k8s.io/apimachinery等项目不同，controller-runtime是kubernetes-sigs 组织下的一个独立项目，而不是一个Kubernetes项目的staging项目。这个项目主要包含用于构建Kubernetes风格控制器的Go语言包集合，主要在Kubebuilder和Operator SDK中被使用。简单地说，controller-runtime是用来引导用户编写最佳实践的 Kubernetes Controllers的。

我们在使用Kubebuilder的时候，其实已经有意无意地在使用controller-runtime了，只是Kubebuilder自动生成了很多代码，所以我们不太能感知到其实用的很多能力都是controller-runtime提供的。

1. Managers

所有的Controller和Webhook最终都是由Managers来运行的，Managers负责Controllers和Webhooks的运行、公共依赖项的设置（pkg/runtime/inject），比如shared caches和clients、管理leader选举（pkg/leaderelection）等。另外，Managers还通过signal handler实现了Pod运行终止时的优雅退出功能（pkg/manager/signals）。

2. Controllers

Controllers（pkg/controller）使用events（pkg/event）来触发调谐请求，可以手动创建Controllers，但是一般都是通过Builder（pkg/builder）来创建的，这样可以简化event源码（pkg/handler），比如Kubernetes资源对象的变更消息，到event处理器之间的关联逻辑编码，或者将一个调谐请求加入所属的队列。Predicates（pkg/predicate）可以被用来过滤哪些event最后会触发调谐过程，其中有一些预置公用代码逻辑用于实现一些进阶场景。

3. Reconcilers

Controllers的逻辑是在Reconcilers中实现的，Reconciler函数的核心逻辑是拿到一个包含name和namespace的对象的调谐请求，然后调谐这个对象，最终返回一个响应或者一个表明是否需要二次调谐的错误。

4. Clients and Caches

Reconcilers使用Clients（pkg/client）来访问API对象，Managers提供的默认Client从本地共享缓存（pkg/cache）中读取数据，直接写到API Server，但是Clients也可以配置成不经过缓存直接和API Server交互。当其他结构化的数据被请求时，缓存中会自动更新监听到的对象。默认单独的Client并不保证缓存的写安全，也不保证创建、查询的一致性。代码不应该假设创建、更新成功的资源能够马上得到更新后的资源。Caches也许有对应的Indexes，它可以通过FieldIndexer（pkg/client）从Managers中获取。Indexes可以被用来快速且简单地通过特定字段检索所有的对象。

5. Schemes

Schemes（pkg/scheme）用来关联Go类型和对应的Kubernetes API类型（Group-Version-Kinds）的。

6. Webhooks

Webhooks（pkg/webhook/admission）也许会被直接实现，但是一般还是使用builder（pkg/webhook/admission/builder）来创建。它们通过被Managers管理的server（pkg/webhook）来运行。

7. Logging and Metrics

Logging（pkg/log）是通过logr（https://godoc.org/github.com/go-logr/logr）日志接口实现的结构化数据，使用Zap（https://go.uber.org/zap, pkg/log/zap）提供了简单的日志配置。也可以使用其他基于logr实现的日志工具作为controller-runtime的日志实现。

Metrics（pkg/metrics）注册到了controller-runtime-specific Prometheus metrics registry中，Manager可以通过HTTP Endpoint来提供Metrics服务。

8. Testing

通过test环境（pkg/envtest）可以简单地给Controllers与Webhooks构建单元测试和集成测试。envtest会自动复制一份ETCD和kube-apiserver到合适的地方，然后提供一个正确的方式来自动连接到API Server。envtest通过一些设计也可以和Ginkgo测试框架一起工作。

6.4 本章小结

本章主要介绍了Operator开发过程中经常会遇到的k8s.io下的依赖项目，希望通过本章的学习能够让读者更熟悉Operator项目的依赖关系，知道自己平时经常使用的一些"工具"是从哪个项目来的，进而遇到问题的时候可以到正确的地方寻找答案。

第 7 章
Operator 开发进阶

本章将详细介绍Operator开发的各种进阶技巧,希望通过本章的学习大家能够对Operator的能力有一个深入的理解,进而在以后的开发实践中能够更加清楚地知道什么样的功能需求适合Operator来做,什么样的功能需求不适合Operator来做。

7.1 进阶项目设计

一般我们需要将一个应用部署到Kubernetes中时,都会选择用Deployment来管理应用,用Service来做服务发现,进一步可能还需要配置 ConfigMap、Ingress、Secret等资源。这么多类型资源的创建、维护、管理工作会变得烦琐且复杂。

我们可以通过Operator在一个应用部署所需的各种资源之上抽象一个Application类型,这个类型里包含必要的一些字段,然后用户只需要创建一个Application,我们通过自定义控制器去完成Application相关的Deployment、Service、ConfigMap等资源的创建和管理工作。

当然,本书主要是为了介绍Operator开发技术,所以这个项目也不真的去实现一个特别完善的Application Operator,而是通过管理Deployment和Service两种资源来演示如何开发Operator。剩下的ConfigMap、Secret等资源如果大家感兴趣,可以自己补充进去。

7.2 准备 application-operator 项目

本节创建一个application-operator项目,同时分析一下这个项目的基础结构。

7.2.1 创建新项目

还记得MyOperatorProjects目录吗?我们之前说过各种演示项目都会放在这个目录下。忘记了也没关系,我们重新创建一个:

```
cd ~
mkdir MyOperatorProjects/
cd MyOperatorProjects/
```

接着在MyOperatorProjects目录下创建我们的学习项目application-operator，但是要注意先确认读者系统的本地没有同名目录，如果有可以选择挪动一下已有的application-operator目录，或者选择换一个路径来创建application-operator项目：

```
# cd ~/MyOperatorProjects/
# mkdir application-operator/
# cd application-operator/
# ls
# kubebuilder init --domain=danielhu.cn \
--repo=github.com/daniel-hutao/application-operator \
--owner Daniel.Hu
Writing kustomize manifests for you to edit...
Writing scaffold for you to edit...
Get controller runtime:
$ go get sigs.k8s.io/controller-runtime@v0.10.0
Update dependencies:
$ go mod tidy
Next: define a resource with:
$ kubebuilder create api
```

这时Kubebuilder就准备好了一个Operator项目骨架，项目名默认是用的当前目录名，也就是application-operator这个目录名。项目名也可以通过--project-name参数来自定义，但是一般没有这个需求，因为很少会有目录名和项目名保持不一致的场景。这里需要知道的是项目名体现在哪些地方，假如中途想修改项目名，直接找到这些配置项，手动调整即可：

1) PROJECT文件中的projectName配置。
2) config/default/kustomization.yaml文件中的namespace配置。
3) config/default/kustomization.yaml文件中的namePrefix配置。

7.2.2 项目基础结构分析

这时Kubebuilder为我们创建了很多文件，接下来详细看一下都有哪些。

1. go.mod

该文件包含项目的基础依赖：

```
module github.com/daniel-hutao/application-operator

1.16

require (
    k8s.io/apimachinery v0.22.1
    k8s.io/client-go v0.22.1
    sigs.k8s.io/controller-runtime v0.10.0
)
```

2. Makefile

该文件中存放的是和开发过程中构建、部署、测试等相关的一系列命令，下面略去了很多不太需要关注的部分：

```
# ...
IMG ?= controller:latest
# ...
manifests: controller-gen
    $(CONTROLLER_GEN) rbac:roleName=manager-role crd webhook paths="./..." output:crd:artifacts:config=config/crd/bases

generate: controller-gen
    $(CONTROLLER_GEN) object:headerFile="hack/boilerplate.go.txt" paths="./..."
# ...
build: generate fmt vet
    go build -o bin/manager main.go

run: manifests generate fmt vet
    go run ./main.go

docker-build: test
    docker build -t ${IMG} .

docker-push:
    docker push ${IMG}
# ...
install: manifests kustomize
    $(KUSTOMIZE) build config/crd | kubectl apply -f -

uninstall: manifests kustomize
    $(KUSTOMIZE) build config/crd | kubectl delete --ignore-not-found=$(ignore-not-found) -f -

deploy: manifests kustomize
    cd config/manager && $(KUSTOMIZE) edit set image controller=${IMG}
    $(KUSTOMIZE) build config/default | kubectl apply -f -

undeploy:
    $(KUSTOMIZE) build config/default | kubectl delete --ignore-not-found=$(ignore-not-found) -f -
```

3. PROJECT

该文件存放的是一些Kubebuilder需要用到的元数据：

```
domain: danielhu.cn
layout:
- go.kubebuilder.io/v3
projectName: application-operator
repo: github.com/daniel-hutao/application-operator
version: "3"
```

4. 部署配置

还有很多与Operator部署相关的配置文件保存在config目录下,主要是运行Controller相关的Kustomize配置以及各种资源配置。

5. main.go

打开application-operator项目的main.go文件,我们来从上往下看。

首先是注释内容:

```
/*
Copyright 2022 Daniel.Hu.

Licensed under the Apache License, Version 2.0 (the "License");
you may not use this file except in compliance with the License.
You may obtain a copy of the License at

    http://www.apache.org/licenses/LICENSE-2.0

Unless required by applicable law or agreed to in writing, software
distributed under the License is distributed on an "AS IS" BASIS,
WITHOUT WARRANTIES OR CONDITIONS OF ANY KIND, either express or implied.
See the License for the specific language governing permissions and
limitations under the License.
*/
```

我们之前提到过每个源文件开头的Copyright声明来自hack/boilerplate.go.txt文件,并且这里会自动加上在kubebuilder init命令中使用--owner=Daniel Hu所指定的作者签名。

接下来是import部分:

```
import (
    "flag"
    "os"

    _ "k8s.io/client-go/plugin/pkg/client/auth"

    "k8s.io/apimachinery/pkg/runtime"
    utilruntime "k8s.io/apimachinery/pkg/util/runtime"
    clientgoscheme "k8s.io/client-go/kubernetes/scheme"
    ctrl "sigs.k8s.io/controller-runtime"
    "sigs.k8s.io/controller-runtime/pkg/healthz"
    "sigs.k8s.io/controller-runtime/pkg/log/zap"
    //+kubebuilder:scaffold:imports
)
```

这里主要是导入了核心的sigs.k8s.io/controller-runtime依赖包,其他包在后面使用的时候再来具体看其作用。

剩下的逻辑其实很简单:

```
func main() {
    var metricsAddr string
    var enableLeaderElection bool
    var probeAddr string
```

```go
        flag.StringVar(&metricsAddr, "metrics-bind-address", ":8080", "The address the
metric endpoint binds to.")
        flag.StringVar(&probeAddr, "health-probe-bind-address", ":8081", "The address
the probe endpoint binds to.")
        flag.BoolVar(&enableLeaderElection, "leader-elect", false,
            "Enable leader election for controller manager. "+
                "Enabling this will ensure there is only one active controller
manager.")
        opts := zap.Options{
            Development: true,
        }
        opts.BindFlags(flag.CommandLine)
        flag.Parse()

        ctrl.SetLogger(zap.New(zap.UseFlagOptions(&opts)))

        mgr, err := ctrl.NewManager(ctrl.GetConfigOrDie(), ctrl.Options{
            Scheme:                 scheme,
            MetricsBindAddress:     metricsAddr,
            Port:                   9443,
            HealthProbeBindAddress: probeAddr,
            LeaderElection:         enableLeaderElection,
            LeaderElectionID:       "ad4dd4db.danielhu.cn",
        })
        if err != nil {
            setupLog.Error(err, "unable to start manager")
            os.Exit(1)
        }

        //+kubebuilder:scaffold:builder

        if err := mgr.AddHealthzCheck("healthz", healthz.Ping); err != nil {
            setupLog.Error(err, "unable to set up health check")
            os.Exit(1)
        }
        if err := mgr.AddReadyzCheck("readyz", healthz.Ping); err != nil {
            setupLog.Error(err, "unable to set up ready check")
            os.Exit(1)
        }

        setupLog.Info("starting manager")
        if err := mgr.Start(ctrl.SetupSignalHandler()); err != nil {
            setupLog.Error(err, "problem running manager")
            os.Exit(1)
        }
    }
```

在main()函数中先设置了一些与metrics相关的标志（flags），接着实例化了一个Manager对象，Manager负责跟踪维护和运行所有的Controllers，同时也设置了共享缓存以及和kube-apiserver通信用的各种Clients。最后在main()函数中通过Start()方法启动了Manager，Manager运行后反过来会启动所有Controller和Webhook。这个Manager会一直运行在后台，直到接收到"优雅停止"信号。

7.3 定义 Application 资源

我们接着来定义和实现API。

7.3.1 添加新 API

创建Application类型和其对应的控制器，其命令是kubebuilder create api：

```
# kubebuilder create api \
--group apps --version v1 --kind Application
Create Resource [y/n]
y
Create Controller [y/n]
y
```

这里我们需要输入两次y，分别在Create Resource [y/n]和Create Controller [y/n]之后。这个命令执行完后，最后几行日志大致是这样的：

```
go get: added sigs.k8s.io/controller-tools v0.7.0
go get: added sigs.k8s.io/structured-merge-diff/v4 v4.1.2
go get: added sigs.k8s.io/yaml v1.2.0
/Users/danielhu/MyOperatorProjects/application-operator/bin/controller-gen object:headerFile="hack/boilerplate.go.txt" paths="./..."
Next: implement your new API and generate the manifests (e.g. CRDs,CRs) with:
$ make manifests
```

再来看api/v1目录，对于一个新的group-version，也就是组和版本都相同的一个资源类型，kubebuilder create api命令会新建一个目录来存放这个group-version。比如这时我们的项目内就多了一个api/v1/目录，该目录也就对应了apps.danielhu.cn/v1这个group-version。还记得我们用的参数--group apps吗？group名会自动加上一开始新建项目时使用--domain所指定的danielhu.cn域名，所以也就变成了apps.danielhu.cn。

然后打开application_types.go文件，有关Copyright部分就不用说了，import部分可以看到只有简单的一行依赖：

```
import (
    metav1 "k8s.io/apimachinery/pkg/apis/meta/v1"
)
```

后面接着是Application类型对应的Spec和Status结构体定义：

```
type ApplicationSpec struct {
    Foo string `json:"foo,omitempty"`
}

type ApplicationStatus struct {
}
```

我们知道Operator的核心逻辑就是不断调谐资源对象的实际状态和期望状态（Spec）保持一

致。这里的Status当然不是严格对应"实际状态",而是观察并记录下来的当前对象最新"状态"。大多数资源对象都有Spec和Status两个部分,但是也有部分资源对象不符合这种模式,比如ConfigMap之类的静态资源对象就不存在着"期望的状态"这一说法。

继续往下可以看到对应Application类型的结构体定义:

```
//+kubebuilder:object:root=true
//+kubebuilder:subresource:status
type Application struct {
    metav1.TypeMeta   `json:",inline"`
    metav1.ObjectMeta `json:"metadata,omitempty"`

    Spec   ApplicationSpec   `json:"spec,omitempty"`
    Status ApplicationStatus `json:"status,omitempty"`
}

//+kubebuilder:object:root=true

type ApplicationList struct {
    metav1.TypeMeta `json:",inline"`
    metav1.ListMeta `json:"metadata,omitempty"`
    Items           []Application `json:"items"`
}
```

Application结构体是Application类型的"根类型",和其他所有的Kubernetes资源类型一样包含TypeMeta和ObjectMeta。TypeMeta中存放的是当前资源的Kind和APIVersion信息,ObjectMeta中存放的是Name、Namespace、Labels和Annotations等信息。

而ApplicationList其实只是简单的一个Application集合类型,其中通过Items存放一组Application,用于List之类的批量操作。

一般情况下,这两个对象都是不需要修改的,我们修改的是前面提到的ApplicationSpec和ApplicationStatus两个结构体。

另外,我们还看到上面的代码中有一行//+kubebuilder:object:root=true这样的特殊注释标记。这个标记主要是被controller-tools识别,然后controller-tools的对象生成器就知道这个标记下面的对象代表一个Kind,接着对象生成器会生成相应的Kind需要的代码,也就是实现runtime.Object接口。换言之,一个结构体要表示一个Kind,必须实现runtime.Object接口。

最后还有一个init()函数:

```
func init() {
    SchemeBuilder.Register(&Application{}, &ApplicationList{})
}
```

7.3.2 自定义新API

接下来定义自己的API,将Application改成需要的样子。如下所示,在ApplicationSpec中添加Deployment和Service属性,类型分别为DeploymentTemplate和ServiceTemplate:

```
// ApplicationSpec defines the desired state of Application
type ApplicationSpec struct {
```

```
    Deployment DeploymentTemplate `json:"deployment,omitempty"`
    Service    ServiceTemplate    `json:"service,omitempty"`
}
```

下一步是定义DeploymentTemplate和ServiceTemplate：

```
type DeploymentTemplate struct {
    appsv1.DeploymentSpec `json:",inline"`
}
type ServiceTemplate struct {
    corev1.ServiceSpec `json:",inline"`
}
```

我们简单地引用Kubernetes原生的DeploymentSpec对象和ServiceSpec对象来构造DeploymentTemplate和ServiceTemplate。

接着就是状态的定义：

```
type ApplicationStatus struct {
    Workflow appsv1.DeploymentStatus `json:"workflow"`
    Network  corev1.ServiceStatus    `json:"network"`
}
```

当然，这里不需要太复杂的逻辑，同样是简单地引用Kubernetes原生的DeploymentStatus对象和ServiceStatus对象来实现状态管理逻辑。

7.4 实现 Application Controller

我们继续来实现控制器的逻辑。

7.4.1 实现主调谐流程

Application资源定义好之后，当然就要开始写控制器的核心调谐逻辑了。打开controllers/application_controller.go源文件，我们可以看到Reconcile方法的骨架。接下来要实现的调谐逻辑大致如图7-1所示。

我们先查看代码，然后对照流程图中的步骤来详细解释：

```
func (r *ApplicationReconciler) Reconcile(ctx context.Context, req ctrl.Request)
(ctrl.Result, error) {
    <-time.NewTicker(100 * time.Millisecond).C
    log := log.FromContext(ctx)

    CounterReconcileApplication += 1
    log.Info("Starting a reconcile", "number", CounterReconcileApplication)

    app := &v1.Application{}
    if err := r.Get(ctx, req.NamespacedName, app); err != nil {
        if errors.IsNotFound(err) {
```

```go
            log.Info("Application not found.")
            return ctrl.Result{}, nil
        }
        log.Error(err, "Failed to get the Application, will requeue after a short time.")
        return ctrl.Result{RequeueAfter: GenericRequeueDuration}, err
    }

    // reconcile sub-resources
    var result ctrl.Result
    var err error

    result, err = r.reconcileDeployment(ctx, app)
    if err != nil {
        log.Error(err, "Failed to reconcile Deployment.")
        return result, err
    }

    result, err = r.reconcileService(ctx, app)
    if err != nil {
        log.Error(err, "Failed to reconcile Service.")
        return result, err
    }

    log.Info("All resources have been reconciled.")
    return ctrl.Result{}, nil
}
```

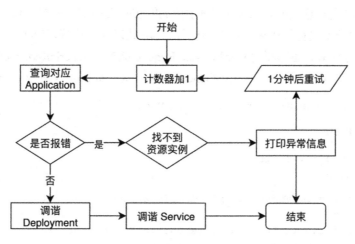

图 7-1 调谐逻辑流程图

1. 计数器

```
<-time.NewTicker(100 * time.Millisecond).C
log := log.FromContext(ctx)

CounterReconcileApplication += 1
log.Info("Starting a reconcile", "number", CounterReconcileApplication)
```

由于调谐过程是并发执行的，也就是说，如果同时创建3个Application类型的资源实例，这时3个事件会同时被处理，日志会比较混乱，所以我们在开头加了一个100毫秒的等待，同时在后面加了一个计数器CounterReconcileApplication，并打印一条日志来输出当前是第几轮调谐。这里不用担心这个数字会溢出，大家如果感兴趣可以计算一下int64有多大，是不是能够让这个程序运行100年也不用担心这里的计数器溢出。CounterReconcileApplication的声明是：

```
var CounterReconcileApplication int64
```

2. 查询对应的Application

```
app := &v1.Application{}
if err := r.Get(ctx, req.NamespacedName, app); err != nil {
    if errors.IsNotFound(err) {
        log.Info("Application not found.")
        return ctrl.Result{}, nil
    }
    log.Error(err, "Failed to get the Application, will requeue after a short time.")
    return ctrl.Result{RequeueAfter: GenericRequeueDuration}, err
}
```

这里先实例化了一个*v1.Application类型的app对象，然后通过r.Get()方法查询触发当前调谐逻辑对应的Application，将其写入app。

接着我们需要处理err != nil的情况，错误分为两种：Application不存在与其他错误。如果是Application不存在，我们的处理是打印一条日志，然后直接返回ctrl.Result{}, nil，也就是意味着"本轮调谐结束"。因为不管是出于什么原因导致Application不存在，比如是被删除了，这时控制器不管进行什么处理都是没有意义的。如果之后不久Application又被创建出来，那么这个调谐过程会被再次触发。所以当前调谐过程只需要直接退出就行了。除此以外的错误，我们需要通过重试来处理，所以除了错误日志打印外，还需要返回ctrl.Result{RequeueAfter: GenericRequeueDuration}, err，也就是在1分钟后再次触发本函数调谐。这里的GenericRequeueDuration定义在文件开头：

```
const GenericRequeueDuration = 1 * time.Minute
```

3. 调谐Deployment

```
var result ctrl.Result
var err error

result, err = r.reconcileDeployment(ctx, app)
if err != nil {
    log.Error(err, "Failed to reconcile Deployment.")
    return result, err
}
```

这里将主要逻辑封装到reconcileDeployment()方法中去实现，从而让主调谐函数看起来可读性更好。当然，reconcileDeployment()方法要做的事情就是完成Deployment资源的调谐过程，然后返回对应的result和error，对于主调谐函数Reconcile()来说，只需要在reconcileDeployment()方法返回的error不等于nil的时候直接返回这个result和error即可。

4. 调谐 Service

```
result, err = r.reconcileService(ctx, app)
if err != nil {
```

```
            log.Error(err, "Failed to reconcile Service.")
            return result, err
    }
```

Service的调谐方式和Deployment基本是一样的，我们同样封装到一个名为reconcileService()的方法中。上面的Deployment调谐过程如果没有任何错误，代码逻辑就会继续走到Service的调谐，最后如果Service的调谐过程没有任何错误，那么主调谐函数的任务就算完成了。所以最后的逻辑是：

```
    log.Info("All resources have been reconciled.")
    return ctrl.Result{}, nil
```

7.4.2 实现 Deployment 调谐流程

在实现控制器的主调谐逻辑时，我们留了两个待实现的子调谐逻辑，也就是reconcileDeployment()方法和reconcileService()方法。接下来，自然是需要实现这两个方法了，我们先从reconcileDeployment()方法开始。

在controllers目录下，与application_controller.go源文件同级目录内创建一个deployment.go源文件来实现Deployment的调谐逻辑。

我们要实现的Deployment调谐逻辑大致如图7-2所示。

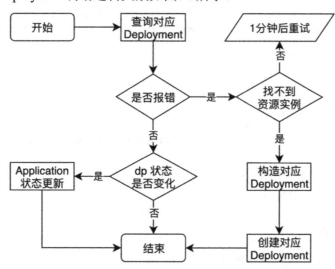

图 7-2　Deployment 调谐逻辑流程图

同样先查看代码，然后对照流程图中的步骤来详细解释：

```go
func (r *ApplicationReconciler) reconcileDeployment(ctx context.Context, app *v1.Application) (ctrl.Result, error) {
    log := log.FromContext(ctx)

    var dp = &appsv1.Deployment{}
    err := r.Get(ctx, types.NamespacedName{
        Namespace: app.Namespace,
        Name:      app.Name,
    }, dp)
```

```go
    if err == nil {
        log.Info("The Deployment has already exist.")
        if reflect.DeepEqual(dp.Status, app.Status.Workflow) {
            return ctrl.Result{}, nil
        }

        app.Status.Workflow = dp.Status
        if err := r.Status().Update(ctx, app); err != nil {
            log.Error(err, "Failed to update Application status")
            return ctrl.Result{RequeueAfter: GenericRequeueDuration}, err
        }
        log.Info("The Application status has been updated.")
        return ctrl.Result{}, nil
    }

    if !errors.IsNotFound(err) {
        log.Error(err, "Failed to get Deployment, will requeue after a short time.")
        return ctrl.Result{RequeueAfter: GenericRequeueDuration}, err
    }

    newDp := &appsv1.Deployment{}
    newDp.SetName(app.Name)
    newDp.SetNamespace(app.Namespace)
    newDp.SetLabels(app.Labels)
    newDp.Spec = app.Spec.Deployment.DeploymentSpec
    newDp.Spec.Template.SetLabels(app.Labels)

    if err := ctrl.SetControllerReference(app, newDp, r.Scheme); err != nil {
        log.Error(err, "Failed to SetControllerReference, will requeue after a short time.")
        return ctrl.Result{RequeueAfter: GenericRequeueDuration}, err
    }

    if err := r.Create(ctx, newDp); err != nil {
        log.Error(err, "Failed to create Deployment, will requeue after a short time.")
        return ctrl.Result{RequeueAfter: GenericRequeueDuration}, err
    }

    log.Info("The Deployment has been created.")
    return ctrl.Result{}, nil
}
```

1. 查询Deployment

```go
log := log.FromContext(ctx)

var dp = &appsv1.Deployment{}
err := r.Get(ctx, types.NamespacedName{
    Namespace: app.Namespace,
    Name:      app.Name,
}, dp)
```

这一步比较常规，先根据Application的Namespace和Name信息来查询对应Deployment是否存在。

2. 没有错误发生时，更新状态

```go
    if err == nil {
```

```
        log.Info("The Deployment has already exist.")
        if reflect.DeepEqual(dp.Status, app.Status.Workflow) {
            return ctrl.Result{}, nil
        }
        app.Status.Workflow = dp.Status
        if err := r.Status().Update(ctx, app); err != nil {
            log.Error(err, "Failed to update Application status")
            return ctrl.Result{RequeueAfter: GenericRequeueDuration}, err
        }
        log.Info("The Application status has been updated.")
        return ctrl.Result{}, nil
    }
```

Deployment的status更新引起的调谐过程被触发。所以接下来判断dp.Status和app.Status.Workflow是否相等，如果不相等，则说明app.Status.Workflow需要更新。

这里大家应该注意到Status更新用的是r.Status().Update()方法。

3. NotFound之外的错误场景

如果错误是NotFound，那么意味着需要创建一个新的Deployment资源实例，这个逻辑很明确，可以预见会有十几到二十行代码。而错误不是NotFound的时候呢？我们只能结束本轮调谐逻辑，选择指定一段时间后去重试。所以我们先写!NotFound的逻辑：

```
    if !errors.IsNotFound(err) {
        log.Error(err, "Failed to get Deployment, will requeue after a short time.")
        return ctrl.Result{RequeueAfter: GenericRequeueDuration}, err
    }
```

4. NotFound的时候

大家应该猜到了，这时要做的就是根据Application资源实例的信息来构造Deployment实例：

```
    newDp := &appsv1.Deployment{}
    newDp.SetName(app.Name)
    newDp.SetNamespace(app.Namespace)
    newDp.SetLabels(app.Labels)
    newDp.Spec = app.Spec.Deployment.DeploymentSpec
    newDp.Spec.Template.SetLabels(app.Labels)

    if err := ctrl.SetControllerReference(app, newDp, r.Scheme); err != nil {
        log.Error(err, "Failed to SetControllerReference, will requeue after a short time.")
        return ctrl.Result{RequeueAfter: GenericRequeueDuration}, err
    }

    if err := r.Create(ctx, newDp); err != nil {
        log.Error(err, "Failed to create Deployment, will requeue after a short time.")
        return ctrl.Result{RequeueAfter: GenericRequeueDuration}, err
    }

    log.Info("The Deployment has been created.")
    return ctrl.Result{}, nil
```

至此，Deployment资源的调谐逻辑也就写好了。

7.4.3 实现 Service 调谐流程

我们继续实现Service的调谐过程。在controllers目录下，即与application_controller.go源文件同级目录内创建一个service.go源文件来实现Service的调谐逻辑。

同样先看图，如图7-3所示，Service的调谐逻辑和Deployment基本一致。

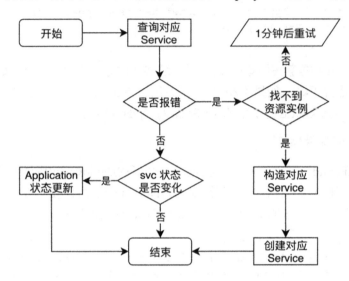

图 7-3　Service 的调谐逻辑流程图

我们还是先查看代码，然后对照流程图中的步骤来详细解释：

```go
func (r *ApplicationReconciler) reconcileService(ctx context.Context, app 
*v1.Application) (ctrl.Result, error) {
    log := log.FromContext(ctx)

    var svc = &corev1.Service{}
    err := r.Get(ctx, types.NamespacedName{
        Namespace: app.Namespace,
        Name:      app.Name,
    }, svc)

    if err == nil {
        log.Info("The Service has already exist.")
        if reflect.DeepEqual(svc.Status, app.Status.Network) {
            return ctrl.Result{}, nil
        }
        app.Status.Network = svc.Status
        if err := r.Status().Update(ctx, app); err != nil {
            log.Error(err, "Failed to update Application status")
            return ctrl.Result{RequeueAfter: GenericRequeueDuration}, err
        }
        log.Info("The Application status has been updated.")
```

```go
        return ctrl.Result{}, nil
    }
    if !errors.IsNotFound(err) {
        log.Error(err, "Failed to get Service, will requeue after a short time.")
        return ctrl.Result{RequeueAfter: GenericRequeueDuration}, err
    }

    newSvc := &corev1.Service{}
    newSvc.SetName(app.Name)
    newSvc.SetNamespace(app.Namespace)
    newSvc.SetLabels(app.Labels)
    newSvc.Spec = app.Spec.Service.ServiceSpec
    newSvc.Spec.Selector = app.Labels

    if err := ctrl.SetControllerReference(app, newSvc, r.Scheme); err != nil {
        log.Error(err, "Failed to SetControllerReference, will requeue after a short time.")
        return ctrl.Result{RequeueAfter: GenericRequeueDuration}, err
    }

    if err := r.Create(ctx, newSvc); err != nil {
        log.Error(err, "Failed to create Service, will requeue after a short time.")
        return ctrl.Result{RequeueAfter: GenericRequeueDuration}, err
    }

    log.Info("The Service has been created.")
    return ctrl.Result{}, nil
}
```

1. 查询Service

```go
log := log.FromContext(ctx)

var svc = &corev1.Service{}
err := r.Get(ctx, types.NamespacedName{
    Namespace: app.Namespace,
    Name:      app.Name,
}, svc)
```

这一步同样是通过Application的Namespace和Name信息来查询资源的,我们这里尝试去Get Service,查看这个Service是否存在。

2. 没有错误发生时,更新状态

```go
if err == nil {
    log.Info("The Service has already exist.")
    if reflect.DeepEqual(svc.Status, app.Status.Network) {
        return ctrl.Result{}, nil
    }

    app.Status.Network = svc.Status
    if err := r.Status().Update(ctx, app); err != nil {
        log.Error(err, "Failed to update Application status")
        return ctrl.Result{RequeueAfter: GenericRequeueDuration}, err
    }
    log.Info("The Application status has been updated.")
```

```
        return ctrl.Result{}, nil
    }
```

这一步是通过比较svc.Status和app.Status.Network的差异来判断Service的Status是否发生了变化，如果有变化就去更新 app.Status。

3. NotFound之外的错误场景

```
if !errors.IsNotFound(err) {
    log.Error(err, "Failed to get Service, will requeue after a short time.")
    return ctrl.Result{RequeueAfter: GenericRequeueDuration}, err
}
```

NotFound之外的错误处理，这里没有新的知识点，我们同样选择在1分钟后重试。

4. NotFound的时候

```
newSvc := &corev1.Service{}
newSvc.SetName(app.Name)
newSvc.SetNamespace(app.Namespace)
newSvc.SetLabels(app.Labels)
newSvc.Spec = app.Spec.Service.ServiceSpec
newSvc.Spec.Selector = app.Labels

if err := ctrl.SetControllerReference(app, newSvc, r.Scheme); err != nil {
    log.Error(err, "Failed to SetControllerReference, will requeue after a short time.")
    return ctrl.Result{RequeueAfter: GenericRequeueDuration}, err
}

if err := r.Create(ctx, newSvc); err != nil {
    log.Error(err, "Failed to create Service, will requeue after a short time.")
    return ctrl.Result{RequeueAfter: GenericRequeueDuration}, err
}
log.Info("The Service has been created.")
return ctrl.Result{}, nil
```

如果Service不存在，就根据Application资源实例的信息来构造Service。至此，Service资源的调谐逻辑也就写好了。

7.4.4　设置 RBAC 权限

我们在前面实现Deployment和Service的调谐逻辑之后，其实还有一个问题没有考虑到，就是这个Operator程序默认是没有权限操作Deployment和Service资源的。当然，我们不需要自己去编写RBAC配置，只需通过几行注释标记代码，工具会自动帮助我们生成相应的配置文件。

回到controllers/application_controller.go文件的Reconcile()方法，可以看到Reconcile()方法上面有这样几行注释：

```
//+kubebuilder:rbac:groups=apps.danielhu.cn,resources=applications,verbs=get;list;watch;create;update;patch;delete
```

```
//+kubebuilder:rbac:groups=apps.danielhu.cn,resources=applications/status,verbs=get;update;patch
//+kubebuilder:rbac:groups=apps.danielhu.cn,resources=applications/finalizers,verbs=update
```

在下方添加操作Deployment和Service相关的注释：

```
//+kubebuilder:rbac:groups=apps,resources=deployments,verbs=get;list;watch;create;update;patch;delete
//+kubebuilder:rbac:groups=apps,resources=deployments/status,verbs=get
//+kubebuilder:rbac:groups=core,resources=services,verbs=get;list;watch;create;update;patch;delete
//+kubebuilder:rbac:groups=core,resources=services/status,verbs=get
```

然后执行如下命令：

```
# make manifests
/Users/danielhu/MyOperatorProjects/application-operator/bin/controller-gen rbac:roleName=manager-role crd webhook paths="./..." output:crd:artifacts:config=config/crd/bases
```

这时打开config/rbac/目录下的role.yaml，可以看到如下配置：

```
---
apiVersion: rbac.authorization.k8s.io/v1
kind: ClusterRole
metadata:
  creationTimestamp: null
  name: manager-role
rules:
- apiGroups:
  - apps
  resources:
  - deployments
  verbs:
  - create
  - delete
  - get
  - list
  - patch
  - update
  - watch
- apiGroups:
  - apps
  resources:
  - deployments/status
  verbs:
  - get
- apiGroups:
  - apps.danielhu.cn
  resources:
  - applications
  verbs:
```

```yaml
    - create
    - delete
    - get
    - list
    - patch
    - update
    - watch
  - apiGroups:
    - apps.danielhu.cn
    resources:
    - applications/finalizers
    verbs:
    - update
  - apiGroups:
    - apps.danielhu.cn
    resources:
    - applications/status
    verbs:
    - get
    - patch
    - update
  - apiGroups:
    - ""
    resources:
    - services
    verbs:
    - create
    - delete
    - get
    - list
    - patch
    - update
    - watch
  - apiGroups:
    - ""
    resources:
    - services/status
    verbs:
    - get
```

可以看到这个ClusterRole中定义了对Deployment和Service等资源的操作权限。但是也有可能会惊奇地发现在自己的环境中执行完make manifests之后，得到的role.yaml文件是这样的：

```yaml
---
apiVersion: rbac.authorization.k8s.io/v1
kind: ClusterRole
metadata:
  creationTimestamp: null
  name: manager-role
rules:
- apiGroups:
  - apps.danielhu.cn
```

```
      resources:
      - applications
      verbs:
      - create
      - delete
      - get
      - list
      - patch
      - update
      - watch
    - apiGroups:
      - apps.danielhu.cn
      resources:
      - applications/finalizers
      verbs:
      - update
    - apiGroups:
      - apps.danielhu.cn
      resources:
      - applications/status
      verbs:
      - get
      - patch
      - update
```

发现区别没有？这个role.yaml中只有Application相关资源的操作权限定义。这是怎么回事呢？其实两种结果的唯一差异就是我们在//+kubebuilder标记和Reconcile方法之间有没有添加空行，也就是前者得到正确的role.yaml时，注释标记是这样的：

```
    //+kubebuilder:rbac:groups=apps.danielhu.cn,resources=applications,verbs=get;list;watch;create;update;patch;delete
    //+kubebuilder:rbac:groups=apps.danielhu.cn,resources=applications/status,verbs=get;update;patch
    //+kubebuilder:rbac:groups=apps.danielhu.cn,resources=applications/finalizers,verbs=update

    //+kubebuilder:rbac:groups=apps,resources=deployments,verbs=get;list;watch;create;update;patch;delete
    //+kubebuilder:rbac:groups=apps,resources=deployments/status,verbs=get
    //+kubebuilder:rbac:groups=core,resources=services,verbs=get;list;watch;create;update;patch;delete
    //+kubebuilder:rbac:groups=core,resources=services/status,verbs=get
    func (r *ApplicationReconciler) Reconcile(ctx context.Context, req ctrl.Request) (ctrl.Result, error) {
        // ...
    }
```

而出错的时候是这样的：

```
    //+kubebuilder:rbac:groups=apps.danielhu.cn,resources=applications,verbs=get;list;watch;create;update;patch;delete
```

```
    //+kubebuilder:rbac:groups=apps.danielhu.cn,resources=applications/status,verbs
=get;update;patch
    //+kubebuilder:rbac:groups=apps.danielhu.cn,resources=applications/finalizers,v
erbs=update
    //+kubebuilder:rbac:groups=apps,resources=deployments,verbs=get;list;watch;crea
te;update;patch;delete
    //+kubebuilder:rbac:groups=apps,resources=deployments/status,verbs=get
    //+kubebuilder:rbac:groups=core,resources=services,verbs=get;list;watch;create;
update;patch;delete
    //+kubebuilder:rbac:groups=core,resources=services/status,verbs=get
    func (r *ApplicationReconciler) Reconcile(ctx context.Context, req ctrl.Request)
(ctrl.Result, error) {
    // ...
    }
```

因为框架默认会生成这样一段注释，而我们平时很容易在看过这段注释后选择将其删除：

```
// Reconcile is part of the main kubernetes reconciliation loop which aims to
// move the current state of the cluster closer to the desired state.
// TODO(user): Modify the Reconcile function to compare the state specified by
// the CronJob object against the actual cluster state, and then
// perform operations to make the cluster state reflect the state specified by
// the user.
//
// For more details, check Reconcile and its Result here:
// - https://pkg.go.dev/sigs.k8s.io/controller-runtime@v0.11.0/pkg/reconcile
```

这段注释被删除之后，就很容易引起上面的RBAC相关注释标记和Reconcile()方法之间丢失空行，接下来controller-gen就不能正确生成role.yaml文件了。

这些工具终究还是不够成熟，虽然很多时候用着不会有问题，但是一旦遇到问题，就很容易被阻塞很长时间，毫无意义地消磨人的技术热情。这里只是提醒大家要注意保留这个空行，算是绕过了这个问题。最优的做法肯定是把这个问题反馈给社区，得到确认后再帮助修复这个问题。或许等这本书写完后空闲下来，笔者会继续跟进这个问题。

7.4.5 过滤调谐事件

我们来思考一下，什么情况下调谐逻辑需要被触发执行一次？

首先Application创建时，肯定是需要执行的，也是整个程序逻辑的第一步。其他的呢？比如Application发生变更时，其实需要根据新的Application中的配置来决定是否需要更新已有的Deployment和Service资源实例。当然，目前我们在调谐Deployment和Service时并没有做得那么完善，只考虑了新建的情况。

再来考虑Application的Status发生变更时应该如何处理，明显我们会根据Deployment和Service资源实例的Status变化来更新Application的Status，这时Status更新了，如果再触发一次调谐，其实这次调谐是没有意义的。

另外，我们也想办法让Deployment和Service的一些变化事件能够选择性地触发调谐逻辑的运行，所以还需要在 **SetupWithManager()** 方法中添加一些程序逻辑。

先看代码：

```go
func (r *ApplicationReconciler) SetupWithManager(mgr ctrl.Manager) error {
    setupLog := ctrl.Log.WithName("setup")

    return ctrl.NewControllerManagedBy(mgr).
        For(&v1.Application{}, builder.WithPredicates(predicate.Funcs{
            CreateFunc: func(event event.CreateEvent) bool {
                return true
            },
            DeleteFunc: func(event event.DeleteEvent) bool {
                setupLog.Info("The Application has been deleted.",
                    "name", event.Object.GetName())
                return false
            },
            UpdateFunc: func(event event.UpdateEvent) bool {
                if event.ObjectNew.GetResourceVersion() == event.ObjectOld.GetResourceVersion() {
                    return false
                }
                if reflect.DeepEqual(event.ObjectNew.(*v1.Application).Spec, event.ObjectOld.(*v1.Application).Spec) {
                    return false
                }
                return true
            },
        })).
        // 1. Deployment
        Owns(&appsv1.Deployment{}, builder.WithPredicates(predicate.Funcs{
            CreateFunc: func(event event.CreateEvent) bool {
                return false
            },
            DeleteFunc: func(event event.DeleteEvent) bool {
                setupLog.Info("The Deployment has been deleted.",
                    "name", event.Object.GetName())
                return true
            },
            UpdateFunc: func(event event.UpdateEvent) bool {
                if event.ObjectNew.GetResourceVersion() == event.ObjectOld.GetResourceVersion() {
                    return false
                }
                if reflect.DeepEqual(event.ObjectNew.(*appsv1.Deployment).Spec, event.ObjectOld.(*appsv1.Deployment).Spec) {
                    return false
                }
                return true
            },
            GenericFunc: nil,
        })).
        // 2. Service
```

```go
        Owns(&corev1.Service{}, builder.WithPredicates(predicate.Funcs{
            CreateFunc: func(event event.CreateEvent) bool {
                return false
            },
            DeleteFunc: func(event event.DeleteEvent) bool {
                setupLog.Info("The Service has been deleted.",
                    "name", event.Object.GetName())
                return true
            },
            UpdateFunc: func(event event.UpdateEvent) bool {
                if event.ObjectNew.GetResourceVersion() == event.ObjectOld.GetResourceVersion() {
                    return false
                }
                if reflect.DeepEqual(event.ObjectNew.(*v1.Application).Spec, event.ObjectOld.(*v1.Application).Spec) {
                    return false
                }
                return true
            },
        })).
        Complete(r)
}
```

这里有三块代码逻辑：Application、Deployment和Service。我们逐个进行分析。

1. Application

对于Application来说，Create事件肯定是无条件处理的；Delete其实不需要做资源清理工作，因为在创建Deployment和Service时都写过类似的代码：

```go
if err := ctrl.SetControllerReference(app, newDp, r.Scheme); err != nil {
    log.Error(err, "Failed to SetControllerReference, will requeue after a short time.")
    return ctrl.Result{RequeueAfter: GenericRequeueDuration}, err
}
```

这些代码的作用就是将当前创建的Deployment资源设置成Application类型的app资源的子资源，这样当对应的Application类型实例被删除时，这个Deployment类型的资源实例就会被系统的垃圾回收系统回收。

Update时需要获取Application中更新的内容，然后将其应用到已创建好的Deployment和Service资源实例中。为了不使得这个项目过于复杂，我们在本书中不展开说明这段程序逻辑。

所以最后For函数应该是这样的：

```go
        For(&v1.Application{}, builder.WithPredicates(predicate.Funcs{
            CreateFunc: func(event event.CreateEvent) bool {
                return true
            },
            DeleteFunc: func(event event.DeleteEvent) bool {
                setupLog.Info("The Application has been deleted.",
                    "name", event.Object.GetName())
```

```
                return false
            },
            UpdateFunc: func(event event.UpdateEvent) bool {
                if event.ObjectNew.GetResourceVersion() ==
event.ObjectOld.GetResourceVersion() {
                    return false
                }
                if reflect.DeepEqual(event.ObjectNew.(*v1.Application).Spec,
event.ObjectOld.(*v1.Application).Spec) {
                    return false
                }
                return true
            },
        }))
```

2. Deployment

Deployment资源对象需要关注的是其变更和删除逻辑。比如用户不小心误删了一个Deployment, 这时控制器应该要能够根据Application配置将其恢复, 或者用户不小心修改了Deployment, 我们也要能够将其恢复回来。对于新增Deployment, 其实不需要做任何处理。比如Application第一次被创建时, 我们在调谐逻辑中会去创建Deployment, 如果这时候再触发一轮调谐, 那么新一轮调谐其实什么事情也做不了。

这里需要注意的是, 如果我们考虑健壮性, 例如用户手动多创建了一个Deployment, 其实控制器需要去检测Deployment的数量, 然后将多余的删除。但是现在主要是为了学习Operator开发, 而不是做一个真正可以上生产环境运行的非常健壮的Operator应用, 所以这类比较复杂的健壮性逻辑先不考虑。

Deployment的事件过滤逻辑就是Create过滤掉, Delete放行, Update选择性放行。于是Owns函数这样编写:

```
        Owns(&appsv1.Deployment{}, builder.WithPredicates(predicate.Funcs{
            CreateFunc: func(event event.CreateEvent) bool {
                return false
            },
            DeleteFunc: func(event event.DeleteEvent) bool {
                setupLog.Info("The Deployment has been deleted.",
                    "name", event.Object.GetName())
                return true
            },
            UpdateFunc: func(event event.UpdateEvent) bool {
                if event.ObjectNew.GetResourceVersion() ==
event.ObjectOld.GetResourceVersion() {
                    return false
                }
                if reflect.DeepEqual(event.ObjectNew.(*appsv1.Deployment).Spec,
event.ObjectOld.(*appsv1.Deployment).Spec) {
                    return false
                }
                return true
            },
```

```
            GenericFunc: nil,
        }))
```

3. Service

Service的事件过滤逻辑其实和Deployment没有多少区别,我们直接看代码:

```
        Owns(&corev1.Service{}, builder.WithPredicates(predicate.Funcs{
            CreateFunc: func(event event.CreateEvent) bool {
                return false
            },
            DeleteFunc: func(event event.DeleteEvent) bool {
                setupLog.Info("The Service has been deleted.",
                    "name", event.Object.GetName())
                return true
            },
            UpdateFunc: func(cvent event.UpdateEvent) bool {
                if event.ObjectNew.GetResourceVersion() ==
event.ObjectOld.GetResourceVersion() {
                    return false
                }
                if reflect.DeepEqual(event.ObjectNew.(*v1.Application).Spec,
event.ObjectOld.(*v1.Application).Spec) {
                    return false
                }
                return true
            },
        }))
```

7.4.6 资源别名

读者可能已经注意到,目前查询Application类型资源时,命令中必须完整输入application(s),也就是大概这样:

```
# kubectl get applications
NAME           AGE
nginx-sample   10s
```

有没有办法像查询DaemonSet时可以用ds来缩写一样,用app来代替application(s)呢?
当然可以,只需要添加这样一行标记在Application结构体之上:

```
//+kubebuilder:resource:path=applications,singular=application,scope=Namespaced,shortName=app
```

然后重新执行make install命令,这时可以通过短命名app来查询Application类型资源实例:

```
# kubectl get app
NAME           AGE
nginx-sample   6m35s
```

7.5 使用 Webhook

Operator除了通过自定义API的方式实现拓展资源管理外，还能通过Webhook方式实现资源访问控制。本节继续学习Webhook。

7.5.1 Kubernetes API 访问控制

我们知道访问Kubernetes API有好几种方式，比如使用kubectl命令、使用client-go之类的开发库、直接通过REST请求等。不管是一个使用kubectl的真人用户，还是一个 ServiceAccount，都可以通过API访问认证，这个过程官网有一张图描述得很直观（https://kubernetes.io/docs/concepts/security/controlling-access/），如图7-4所示。

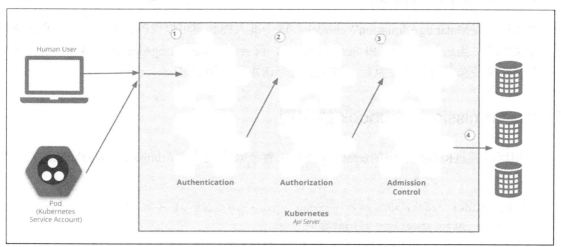

图 7-4　Kubernetes API 访问控制流程

当一个访问请求发送到API Server的时候，会依次经过认证、鉴权、准入控制三个主要的过程。本节主要学习的Admission Webhook就是这里提到的"准入控制"的范畴。

准入控制（Admission Control）模块能够实现更改一个请求的内容或者决定是否拒绝一个请求的功能。准入控制主要是在一个对象发生变更时生效，变更包括创建、更新、删除等动作，也就是不包含查询动作。如果配置了多个准入控制模块，那么这些模块是按顺序工作的。

关于拒绝请求这个能力，一个请求在多个准入控制模块中有一个模块拒绝，这个请求就会被拒绝，这和认证或者鉴权模块明显不一样。而更改一个请求内容的能力，主要用于给一些请求字段设置默认值。

准入控制器基本都是在kube-apiserver中实现的，所以它们的启用也是通过在kube-apiserver的启动参数上添加相应配置，比如：

```
kube-apiserver --enable-admission-plugins=NamespaceLifecycle,LimitRanger ...
```

大家可以在https://kubernetes.io/docs/reference/access-authn-authz/admission-controllers/#what-does-each-admission-controller-do看到目前有哪些准入控制器以及它们的作用。这里的多数准入

控制器只能决定它们的启用或者禁用，除了这类在kube-apiserver内部实现的准入控制器外，我们可以看到有两个特殊的准入控制器：ValidatingAdmissionWebhook和MutatingAdmissionWebhook。这是Kubernetes提供的一种拓展机制，让我们能够通过Webhook的方式独立于kube-apiserver运行自己的准入控制逻辑。

7.5.2 Admission Webhook 介绍

顾名思义，Admission Webhook是一个HTTP回调钩子，可以用来接收"准入请求"，然后对这个请求做相应的逻辑处理。

Admission Webhook有两种：

- ValidatingAdmissionWebhook。
- MutatingAdmissionWebhook。

先执行的是MutatingAdmissionWebhook，这个准入控制器可以修改请求对象，主要用来注入自定义字段；当这个对象被API Server校验时，就会回调ValidatingAdmissionWebhook，然后相应的自定义校验策略就会被执行，以决定这个请求能否被通过。

7.5.3 Admission Webhook 的实现

我们可以通过Kubebuilder的create webhook命令来生成实现Admission Webhook的代码脚手架：

```
# kubebuilder create webhook --group apps --version v1 --kind Application
--defaulting --programmatic-validation                    1 ↵
    Writing kustomize manifests for you to edit...
    Writing scaffold for you to edit...
    api/v1/application_webhook.go
    Update dependencies:
    $ go mod tidy
    Running make:
    $ make generate
    /Users/danielhu/MyOperatorProjects/application-operator/bin/controller-gen
object:headerFile="hack/boilerplate.go.txt" paths="./..."
    Next: implement your new Webhook and generate the manifests with:
    $ make manifests
```

这个命令执行完成后，可以看到项目内多了文件。打开api/v1/application_webhook.go源文件，可以看到里面有一个Default()方法。在Default()方法中就可以完成MutatingAdmissionWebhook的相关逻辑。

1. 实现MutatingAdmissionWebhook

我们以Replicas默认值注入为例，比如用户提交的Application配置中没有给出Replicas的大小，那么注入一个默认值3，代码如下：

```go
func (r *Application) Default() {
    applicationlog.Info("default", "name", r.Name)

    if r.Spec.Deployment.Replicas == nil {
        r.Spec.Deployment.Replicas = new(int32)
        *r.Spec.Deployment.Replicas = 3
    }
}
```

2. 实现ValidatingAdmissionWebhook

在application_webhook.go源文件中继续往后看,可以发现有3个Validate×××()方法,分别是ValidateCreate、ValidateUpdate和ValidateDelete。顾名思义,这几个Validate方法的触发条件分别是相应对象在创建、更新、删除的时候。

删除时不需要做什么校验逻辑,而创建和更新的校验逻辑几乎是一样的,所以我们将创建和更新时所需的校验逻辑封装一下,编写一个validateApplication()方法:

```go
func (r *Application) validateApplication() error {
    if *r.Spec.Deployment.Replicas > 10 {
        return fmt.Errorf("replicas too many error")
    }
    return nil
}
```

这里我们简单地校验Replicas是不是设置得过大了,其他业务逻辑也是类似的校验方法,如果觉得条件不满足,就返回一个error,反之返回nil就行。

接下来,在几个Validate×××()方法中调用这个validateApplication()方法:

```go
func (r *Application) ValidateCreate() error {
    applicationlog.Info("validate create", "name", r.Name)

    return r.validateApplication()
}

func (r *Application) ValidateUpdate(old runtime.Object) error {
    applicationlog.Info("validate update", "name", r.Name)

    return r.validateApplication()
}

func (r *Application) ValidateDelete() error {
    applicationlog.Info("validate delete", "name", r.Name)
    return nil
}
```

这时如果想在本地运行测试Webhook,默认需要准备证书,放到/tmp/k8s-webhook-server/serving-certs/tls.{crt,key}中,然后执行make run命令。

如果真的这样去执行,大概率会遇到这个错误(可能读者不会遇到,因为这本书出版的时候Kubebuilder很可能已经修复这个问题):

```
api/v1/application_webhook.go:40:5: cannot use &Application{} (type *Application) as type admission.Defaulter in assignment:
```

```
                *Application does not implement admission.Defaulter (missing DeepCopyObject
method)
        api/v1/application_webhook.go:55:5: cannot use &Application{} (type *Application)
as type admission.Validator in assignment:
                *Application does not implement admission.Validator (missing DeepCopyObject
method)
```

怎么解决呢？大家可以在这个issue中找到一些线索：https://github.com/kubernetes-sigs/kubebuilder/issues/1558#issuecomment-668687181。最后Application的注解大致是这样的：

```
//+kubebuilder:subresource:status
//+kubebuilder:resource:path=applications,singular=application,scope=Namespaced,shortName=app
//+kubebuilder:object:root=true
type Application struct {
    metav1.TypeMeta   `json:",inline"`
    metav1.ObjectMeta `json:"metadata,omitempty"`

    Spec   ApplicationSpec   `json:"spec,omitempty"`
    Status ApplicationStatus `json:"status,omitempty"`
}
```

7.5.4　cert-manager 部署

在部署Webhook之前需要先安装cert-manager，用来实现证书签发功能。关于cert-manager的详细介绍大家可以参考官方文档：https://cert-manager.io/docs/，本节我们只介绍怎么部署cert-manager。

cert-manager提供了helm Chart包方式部署，我们来一步一步以helm的方式部署cert-manager：

```
# helm repo add jetstack https://charts.jetstack.io
"jetstack" has been added to your repositories
# helm repo update
Hang tight while we grab the latest from your chart repositories...
...Successfully got an update from the "jetstack" chart repository
Update Complete. ⎈Happy Helming!⎈
# helm search repo jetstack
NAME                                       CHART VERSION   APP VERSION   DESCRIPTION
jetstack/cert-manager                      v1.8.0          v1.8.0        A Helm chart for cert-manager
jetstack/cert-manager-approver-policy      v0.3.0          v0.3.0        A Helm chart for cert-manager-approver-policy
jetstack/cert-manager-csi-driver           v0.2.1          v0.2.1        A Helm chart for cert-manager-csi-driver
jetstack/cert-manager-csi-driver-spiffe    v0.1.0          v0.1.0        A Helm chart for cert-manager-csi-driver-spiffe
jetstack/cert-manager-istio-csr            v0.4.2          v0.4.0        istio-csr enables the use of cert-manager for i...
jetstack/cert-manager-trust                v0.1.1          v0.1.0        A Helm chart for cert-manager-trust
```

可以看到cert-manager对应的Chart名字是jetstack/cert-manager，我们安装这个Chart：

```
# helm install \
  cert-manager jetstack/cert-manager \
  --namespace cert-manager \
  --create-namespace \
  --version v1.8.0 \
  --set installCRDs=true
NAME: cert-manager
LAST DEPLOYED: Sun May  1 10:46:47 2022
NAMESPACE: cert-manager
STATUS: deployed
REVISION: 1
TEST SUITE: None
NOTES:
cert-manager v1.8.0 has been deployed successfully!

In order to begin issuing certificates, you will need to set up a ClusterIssuer
or Issuer resource (for example, by creating a 'letsencrypt-staging' issuer).

More information on the different types of issuers and how to configure them
can be found in our documentation:

https://cert-manager.io/docs/configuration/

For information on how to configure cert-manager to automatically provision
Certificates for Ingress resources, take a look at the `ingress-shim`
documentation:

https://cert-manager.io/docs/usage/ingress/
```

这样cert-manager就部署完成了，我们可以检查创建的相关Pod是不是都正常运行：

```
# kgpo -n cert-manager
NAME                                       READY   STATUS    RESTARTS   AGE
cert-manager-6bbf595697-jbgq9              1/1     Running   0          5m55s
cert-manager-cainjector-6bc9d758b-ldqxc    1/1     Running   0          5m55s
cert-manager-webhook-586d45d5ff-9w8h8      1/1     Running   0          5m55s
```

7.5.5　Webhook 部署运行

我们已经准备好了Webhook代码，接着就部署到环境中来看一下运行结果。

1. 构建并推送镜像

我们执行以下两行命令来构建镜像，并把镜像加载到kind集群中：

```
make docker-build IMG=application-operator:v0.1
kind load docker-image application-operator:v0.1 --name dev
```

2. 部署CRD

CRD的部署很简单，执行以下命令：

```
# make install
/Users/danielhu/MyOperatorProjects/application-operator/bin/controller-gen
rbac:roleName=manager-role crd webhook paths="./..."
output:crd:artifacts:config=config/crd/bases
    /Users/danielhu/MyOperatorProjects/application-operator/bin/kustomize build
config/crd | kubectl apply -f -
    customresourcedefinition.apiextensions.k8s.io/applications.apps.danielhu.cn
configured
```

3. 证书相关配置

我们前面部署了cert-manager，要使用cert-manager还需要做一些配置。首先config/default/kustomization.yaml文件需要做一些调整，打开几行注释内容，最后看起来应该是这样的：

```yaml
namespace: application-operator-system
namePrefix: application-operator-
bases:
- ../crd
- ../rbac
- ../manager
- ../webhook
- ../certmanager
patchesStrategicMerge:
- manager_auth_proxy_patch.yaml
- manager_webhook_patch.yaml
- webhookcainjection_patch.yaml
vars:
- name: CERTIFICATE_NAMESPACE # namespace of the certificate CR
  objref:
    kind: Certificate
    group: cert-manager.io
    version: v1
    name: serving-cert # this name should match the one in certificate.yaml
  fieldref:
    fieldpath: metadata.namespace
- name: CERTIFICATE_NAME
  objref:
    kind: Certificate
    group: cert-manager.io
    version: v1
    name: serving-cert # this name should match the one in certificate.yaml
- name: SERVICE_NAMESPACE # namespace of the service
  objref:
    kind: Service
    version: v1
    name: webhook-service
  fieldref:
    fieldpath: metadata.namespace
- name: SERVICE_NAME
  objref:
    kind: Service
```

```
  version: v1
  name: webhook-service
```

接着还需要调整config/crd/kustomization.yaml文件,修改如下:

```
resources:
- bases/apps.danielhu.cn_applications.yaml
#+kubebuilder:scaffold:crdkustomizeresource

patchesStrategicMerge:
- patches/webhook_in_applications.yaml

#+kubebuilder:scaffold:crdkustomizewebhookpatch
- patches/cainjection_in_applications.yaml

#+kubebuilder:scaffold:crdkustomizecainjectionpatch
configurations:
- kustomizeconfig.yaml
```

4. 部署控制器

接下来可以部署控制器,执行以下命令:

```
# make deploy IMG=application-operator:v0.1
/Users/danielhu/MyOperatorProjects/application-operator/bin/controller-gen rbac:roleName=manager-role crd webhook paths="./..." output:crd:artifacts:config=config/crd/bases
    cd config/manager && /Users/danielhu/MyOperatorProjects/application-operator/bin/kustomize edit set image controller=application-operator:v0.1
    /Users/danielhu/MyOperatorProjects/application-operator/bin/kustomize build config/default | kubectl apply -f -
    namespace/application-operator-system created
    customresourcedefinition.apiextensions.k8s.io/applications.apps.danielhu.cn configured
    // ...
    deployment.apps/application-operator-controller-manager created
```

5. 查看结果

最后可以查看Pod是否正常运行:

```
# kubectl get pod -n application-operator-system
NAME                                                        READY   STATUS    RESTARTS   AGE
application-operator-controller-manager-8bc45f9d6-h9m8q     2/2     Running   0          47m
```

7.5.6 Webhook测试

我们准备一个YAML配置:

```
apiVersion: apps.danielhu.cn/v1
kind: Application
metadata:
```

```
      name: nginx-sample
      namespace: default
      labels:
        app: nginx
    spec:
      deployment:
        replicas: 11
        selector:
          matchLabels:
            app: nginx
        template:
          metadata:
            labels:
              app: nginx
          spec:
            containers:
              - name: nginx
                image: nginx:1.14.2
                ports:
                  - containerPort: 80
      service:
        type: NodePort
        ports:
          - port: 80
            targetPort: 80
            nodePort: 30080
```

注意这里的replicas给了11，然后应用一下看看Validator是否生效：

```
# kubectl apply -f apps_v1_application.yaml
Error from server (replicas too many error): error when applying patch:
{"metadata":{"annotations":{"kubectl.kubernetes.io/last-applied-configuration":
"{\"apiVersion\":\"apps.danielhu.cn/v1\",\"kind\":\"Application\",\"metadata\":{\"a
nnotations\":{},\"labels\":{\"app\":\"nginx\"},\"name\":\"nginx-sample\",\"namespac
e\":\"default\"},\"spec\":{\"deployment\":{\"replicas\":11,\"selector\":{\"matchLab
els\":{\"app\":\"nginx\"}},\"template\":{\"metadata\":{\"labels\":{\"app\":\"nginx\
"}},\"spec\":{\"containers\":[{\"image\":\"nginx:1.14.2\",\"name\":\"nginx\",\"port
s\":[{\"containerPort\":80}]}]}}},\"service\":{\"ports\":[{\"nodePort\":30080,\"por
t\":80,\"targetPort\":80}],\"type\":\"NodePort\"}}}\n"}},"spec":{"deployment":{"rep
licas":11,"template":{"metadata":{"labels":{"app":"nginx"}},"spec":{"containers":[{
"image":"nginx:1.14.2","name":"nginx","ports":[{"containerPort":80}]}]}}},"service"
:{"ports":[{"nodePort":30080,"port":80,"targetPort":80}]}}}
to:
Resource: "apps.danielhu.cn/v1, Resource=applications", GroupVersionKind:
"apps.danielhu.cn/v1, Kind=Application"
Name: "nginx-sample", Namespace: "default"
for: "apps_v1_application.yaml": admission webhook "vapplication.kb.io" denied the
request: replicas too many error
```

符合预期，我们得到了一个replicas too many error错误。

接着将Replicas删除，使用同样的方式可以验证Defaulter能不能正常工作。结果是在不设置Replicas的情况下，Replicas默认值会变成3。大家可以自行验证，这里不再赘述。

7.6　API 多版本支持

一般情况下，我们开发一个新项目，它的API是会经常变更的，不管一开始考虑得多么详细，都避免不了迭代的过程中去修改API定义。过了一段时间后，API会趋于稳定，在达到稳定版本之后，可能我们才正式发布V1.0版本。当然，这个稳定版的API就不应该再变化了，如果过了半年想增强一下这个API，可能需要发布V2.0版本，这时V1.0版本还是要能够继续正常工作。本节看一下Operator中是如何支持多版本API的。

7.6.1　实现 V2 版本 API

我们先通过kubebuilder命令添加一个V2版本的API：

```
# kubebuilder create api \
    --group apps \
    --version v2 \
    --kind Application
Create Resource [y/n]
y
Create Controller [y/n]
n
Writing kustomize manifests for you to edit...
Writing scaffold for you to edit...
api/v2/application_types.go
Update dependencies:
$ go mod tidy
Running make:
$ make generate
/Users/danielhu/MyOperatorProjects/application-operator/bin/controller-gen object:headerFile="hack/boilerplate.go.txt" paths="./..."
Next: implement your new API and generate the manifests (e.g. CRDs,CRs) with:
$ make manifests
```

需要注意的是，这里不要创建Controller。我们通过Git查看目前发生变更的文件有哪些，如图7-5所示。

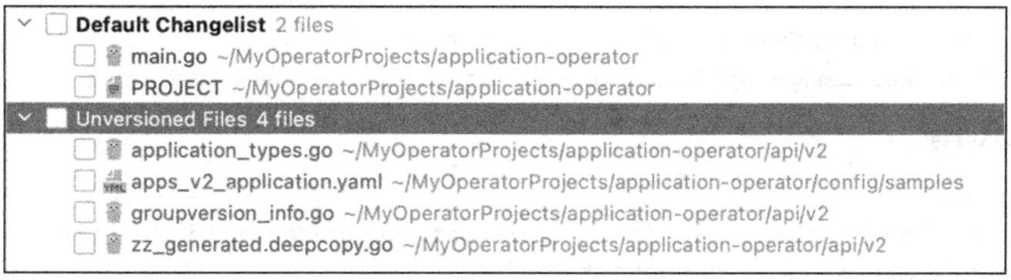

图 7-5　发生变更的文件

可以看到api目录下多了一个v2目录，另外main.go和PROJECT发生了变化。大家可以逐个点开查看具体的变更内容，这里就不赘述这个变化了，因为新增的内容和V1版本其实是对称一致的。

V2版本只用于演示多版本API，所以这里不用添加太多的功能。前面在Application类型的Spec中定义了一个Deployment字段，假如想要将其改成更通用的Workflow，我们在api/v2目录下的application_types.go文件中实现和V1版本完全一样的代码逻辑（除了package v2这一行有差异），然后把ApplicationSpec结构体修改如下：

```go
type ApplicationSpec struct {
    Workflow v1.DeploymentTemplate `json:"workflow,omitempty"`
    Service  v1.ServiceTemplate    `json:"service,omitempty"`
}
```

至此，代码就改完了。但是有了多个版本的API，我们在API Server中却只能指定持久化一个版本，这里选择持久化V1版本，所以需要在V1版本的Application上增加一行注解"//+kubebuilder:storageversion"，最后的代码如下：

```go
//+kubebuilder:subresource:status
//+kubebuilder:resource:path=applications,singular=application,scope=Namespaced,shortName=app
//+kubebuilder:storageversion
//+kubebuilder:object:root=true
type Application struct {
    metav1.TypeMeta   `json:",inline"`
    metav1.ObjectMeta `json:"metadata,omitempty"`

    Spec   ApplicationSpec   `json:"spec,omitempty"`
    Status ApplicationStatus `json:"status,omitempty"`
}
```

至此，新版本类型的实现就算完成了。

这时应该添加一个Webhook用来接收API Server的conversion回调请求（参考https://book.kubebuilder.io/multiversion-tutorial/webhooks.html）。7.5.6节其实已经配置过Webhook了，所以这里什么也不需要做，已经可以开始部署测试多版本API了。

7.6.2 多版本 API 部署测试

接下来测试多版本API是否和预期的一致，我们验证一下。

1. 构建镜像并且推给kind集群

```
make docker-build IMG=application-operator:v0.2
kind load docker-image application-operator:v0.2 --name dev
```

2. 部署CRD等

```
# make install
/Users/danielhu/MyOperatorProjects/application-operator/bin/controller-gen rbac:roleName=manager-role crd webhook paths="./..." output:crd:artifacts:config=config/crd/bases
```

```
/Users/danielhu/MyOperatorProjects/application-operator/bin/kustomize build
config/crd | kubectl apply -f -
The CustomResourceDefinition "applications.apps.danielhu.cn" is invalid:
metadata.annotations: Too long: must have at most 262144 bytes
make: *** [install] Error 1
```

又遇到bug了,开源项目就是经常会遇到这类情况,很可能跑着跑着就掉坑里了,这会给初学者带来很大的困扰。我们可以在这个地址:https://github.com/kubernetes-sigs/kubebuilder/issues/1140找到关于这个bug的描述。

要绕过这个问题,只需要在Makefile中将install的命令改成如下代码就行(下面的deploy同理也需要将apply改成create):

```
.PHONY: install
install: manifests kustomize ## Install CRDs into the K8s cluster specified in ~/.kube/config.
    $(KUSTOMIZE) build config/crd | kubectl create -f -
```

然后重新执行install:

```
# make install
/Users/danielhu/MyOperatorProjects/application-operator/bin/controller-gen
rbac:roleName=manager-role crd webhook paths="./..."
output:crd:artifacts:config=config/crd/bases
/Users/danielhu/MyOperatorProjects/application-operator/bin/kustomize build
config/crd | kubectl create -f -
customresourcedefinition.apiextensions.k8s.io/applications.apps.danielhu.cn
created
```

3. 部署Operator

```
# make deploy IMG=application-operator:v0.2
/Users/danielhu/MyOperatorProjects/application-operator/bin/controller-gen
rbac:roleName=manager-role crd webhook paths="./..."
output:crd:artifacts:config=config/crd/bases
    cd config/manager &&
/Users/danielhu/MyOperatorProjects/application-operator/bin/kustomize edit set image
controller=application-operator:v0.2
/Users/danielhu/MyOperatorProjects/application-operator/bin/kustomize build
config/default | kubectl create -f -
    namespace/application-operator-system created
    // ...
    validatingwebhookconfiguration.admissionregistration.k8s.io/application-operator-validating-webhook-configuration created
```

4. 结果检查

这时可以查看Operator是否正常运行:

```
# kubectl get pod -n application-operator-system
NAME                                                         READY   STATUS    RESTARTS   AGE
application-operator-controller-manager-57dbfb8479-r5wgz     2/2     Running   0          68s
```

5. 部署V2版本资源

我们准备一个新的资源YAML配置：

```yaml
apiVersion: apps.danielhu.cn/v2
kind: Application
metadata:
  name: nginx-sample
  namespace: default
  labels:
    app: nginx
spec:
  workflow:
    replicas: 1
    selector:
      matchLabels:
        app: nginx
    template:
      metadata:
        labels:
          app: nginx
      spec:
        containers:
          - name: nginx
            image: nginx:1.14.2
            ports:
              - containerPort: 80
  service:
    type: NodePort
    ports:
      - port: 80
        targetPort: 80
        nodePort: 30080
```

注意到这里用了apiVersion: apps.danielhu.cn/v2，然后通过kubectl apply这个YAML文件就完成了资源的创建：

```
# kubectl apply -f apps_v2_application.yaml
application.apps.danielhu.cn/nginx-sample created
```

6. 通过V2版本查询

我们通过V2版本创建的资源，当然可以查询到该版本的资源：

```
# kubectl get applications.v2.apps.danielhu.cn -o yaml
apiVersion: v1
items:
- apiVersion: apps.danielhu.cn/v2
  kind: Application
  // ...
  spec:
    service:
```

```
      // ...
      workflow:
        replicas: 1
        // ...
```

7. 通过V1版本查询

大家应该记得，其实是通过V1版本存储的这个资源，所以通过V1版本也能查到这个资源：

```
# kubectl get applications.v1.apps.danielhu.cn -o yaml
apiVersion: v1
items:
- apiVersion: apps.danielhu.cn/v2
  kind: Application
  // ...
  spec:
    deployment:
      replicas: 1
      // ...
    service:
      // ...
```

可以看到，多版本API正常工作了，通过两个版本都可以查询到资源。其实这也说明Convert函数在正常工作。

7.7 API 分组支持

有时我们会在一个Operator项目中实现多个控制器来管理不同的API资源组。比如如果要实现一个ai-operator项目，其中可能包含模型训练相关的控制器trainjob-controller和推理服务相关的控制器application-controller。那么如何将API分别放到apps组和batch组中呢？我们继续来学习API分组怎么实现。

首先通过Kubebuilder做一些简单的工作：

```
kubebuilder edit --multigroup=true
```

这个命令的能力其实非常有限，它只是在Dockerfile文件中将api/目录变成了apis/目录，然后在PROJECT文件中加了一行multigroup: true。

接着还有一些手动工作需要完成：

```
cd ~/MyOperatorProjects/application-operator
mdkir -p apis/apps
mv api/v* apis/apps/
rm -rf api/
```

这时旧版的API就被挪动到合适的目录了。接着还需要挪动controllers目录内的控制器相关源文件：

```
mkdir -p controllers/apps
mv controllers/*.go controllers/apps/
```

目录结构移动好了。接下来就是对应源文件内的import部分引用的更新了。如果大家是用IDE做的重构，这些引用应该会自动更新。import的更新非常简单，这里不再赘述。

还有一个细节就是suite_test.go中和envtest相关的代码要做一个小更新，因为将这个源文件放到了更深一层的目录中，所以这行代码如下：

```
CRDDirectoryPaths:     []string{filepath.Join("..", "config", "crd", "bases")},
```

需要更新为：

```
CRDDirectoryPaths:     []string{filepath.Join("..", "..", "config", "crd", "bases")},
```

到这里就完成了API分组支持的配置。记得PROJECT文件中多出来的一行multigroup: true配置吗？这行配置会告诉Kubebuilder后面新增的API不要放到api/<version>目录下，而是放到apis/<group>/<version>目录下；另外控制器的代码也不再放到controllers目录下，而是放到controllers/<group>目录下。

7.8 本章小结

本章我们通过实现一个调谐Deployment和Service资源的Application控制器来介绍各种Operator开发相关的进阶知识。另外，我们也学习了如何实现多个分组的API、如何实现API多版本、如何使用Webhook等知识点。希望通过本章的学习，大家能够对Operator开发有更深的认识。

第 8 章 Deployment Controller 源码分析

我们在深度使用Operator模式进行开发时，虽然可以借助Kubebuilder和Operator SDK等工具较好地屏蔽底层细节，让我们专注于自身业务逻辑，但是不清楚底层原理会让我们在编码过程中心里没底，比如：自定义控制器重启时会重新收到所有相关资源的Event吗？我们调谐的子资源是Deployment时相关Pod的变更会触发调谐逻辑吗？很多类似的细节问题会不停跳出来，让你对自己编写的代码没有信心。所以我们想要真正深入Operator开发，写出高效、健壮的代码，就必须多阅读一些相关组件的实现原理和源码，从而对自己开发的自定义控制器行为能够知根知底，胸有成竹。

Deployment是常用的Kubernetes原生工作负载资源之一，刚开始尝试使用Kubernetes时大概率就是从运行一个Deployment类型的工作负载开始的。本章将从Deployment的特性介绍、源码分析等方面深度剖析Deployment资源及其背后的Deployment控制器的工作原理。

8.1 Deployment 功能分析

Deployment的基础特性大家肯定都熟悉，所以本节不赘述Deployment的所有功能细节，而是从滚动更新等不算基础的特性入手，看一下Deployment支持哪些特性，为后面分析源码做准备。

8.1.1 Deployment 基础知识

我们先创建一个简单的Deployment，然后查看其中的一些细节。以运行nginx为例，可以通过Deployment来启动一个具有3个副本的nginx负载，先准备YAML文件：

```
cat <<EOF >./nginx-deployment.yaml
apiVersion: apps/v1
kind: Deployment
metadata:
  name: nginx
  labels:
    app: nginx
spec:
  replicas: 3
  selector:
    matchLabels:
      app: nginx
  template:
    metadata:
      labels:
        app: nginx
    spec:
      containers:
      - name: nginx
        image: nginx:1.14.2
        ports:
        - containerPort: 80
EOF
```

然后通过kubectl来部署这个Deployment：

```
# kubectl create -f nginx-deployment.yaml
deployment.apps/nginx created
# kubectl get deploy
NAME     READY   UP-TO-DATE   AVAILABLE   AGE
nginx    1/3     3            1           3s
# kubectl get deploy
NAME     READY   UP-TO-DATE   AVAILABLE   AGE
nginx    3/3     3            3           10s
```

等不了多久，就可以通过kubectl get deploy命令看到所有Pod都建立起来了。这里关注该命令输出字段的含义（NAME和AGE就不用说了）：

- UP-TO-DATE：已经更新到期望状态的副本数。
- AVAILABLE：已经可以提供服务的副本数。
- READY：可以提供服务的副本数/期望副本数。

继续查询一下ReplicaSet：

```
# kubectl get rs --selector=app=nginx
NAME                DESIRED   CURRENT   READY   AGE
nginx-66b6c48dd5    3         3         3       3m54s
```

创建了Deployment之后，可以看到集群中多了一个ReplicaSet资源，也就是说Deployment管理的其实是ReplicaSet，而不是直接管理Pod。我们继续看ReplicaSet的配置来验证这个想法：

```
# kubectl get rs nginx-66b6c48dd5 -o yaml
```

```
apiVersion: apps/v1
kind: ReplicaSet
// ...
  ownerReferences:
  - apiVersion: apps/v1
    blockOwnerDeletion: true
    controller: true
    kind: Deployment
    name: nginx
    uid: 97736b65-0171-4916-bb18-feccc343ac14
  resourceVersion: "1099157"
  uid: 83ac5660-28eb-4d40-beb1-cb5ceb6928b6
// ...
```

这里可以看到ReplicaSet属于Deployment类型的nginx资源。用同样的方法可以看到对应的Pod是ReplicaSet管理的。

到这里，我们可以猜一下Deployment Controller的实现原理，大概可以想到其通过管理ReplicaSet的生命周期，借助ReplicaSet Controller提供的能力间接完成Pod生命周期的管理。另外，它可以通过创建多个ReplicaSet资源，控制其副本数来实现滚动更新和回滚等操作。这样Deployment Controller的实现逻辑就相对"高层"了。

8.1.2 Deployment 的滚动更新和回滚

我们继续看一下Deployment的滚动更新过程，通过kubectl set命令来更新镜像：

```
# kubectl set image deployment/nginx nginx=nginx:1.16.1
deployment.apps/nginx image updated
```

然后查看一下Event信息：

```
# kubectl describe deploy nginx
// ...

Events:
  Type    Reason             Age   From                   Message
  ----    ------             ----  ----                   -------
  Normal  ScalingReplicaSet  26m   deployment-controller  Scaled up replica set nginx-66b6c48dd5 to 3
  Normal  ScalingReplicaSet  88s   deployment-controller  Scaled up replica set nginx-559d658b74 to 1
  Normal  ScalingReplicaSet  87s   deployment-controller  Scaled down replica set nginx-66b6c48dd5 to 2
  Normal  ScalingReplicaSet  87s   deployment-controller  Scaled up replica set nginx-559d658b74 to 2
  Normal  ScalingReplicaSet  86s   deployment-controller  Scaled down replica set nginx-66b6c48dd5 to 1
  Normal  ScalingReplicaSet  86s   deployment-controller  Scaled up replica set nginx-559d658b74 to 3
  Normal  ScalingReplicaSet  84s   deployment-controller  Scaled down replica set nginx-66b6c48dd5 to 0
```

从Event中可以看到，deployment-controller通过调整ReplicaSet资源nginx-66b6c48dd5和nginx-559d658b74的副本数完成了这次滚动更新。先看这两个ReplicaSet：

```
# kubectl get rs --selector=app=nginx
NAME                DESIRED   CURRENT   READY   AGE
nginx-559d658b74    3         3         3       134m
nginx-66b6c48dd5    0         0         0       159m
```

可以看到这时新增了一个nginx-559d658b74，副本数是3，同时旧的nginx-66b6c48dd5 变成了0副本。这个过程大概是这样的：

1）nginx-66b6c48dd5 to 3 / replica set nginx-559d658b74 to 1：新rs增加一个副本到1，合计4个副本。

2）Scaled down replica set nginx-66b6c48dd5 to 2：旧rs减少一个副本到2，合计3个副本。

3）Scaled up replica set nginx-559d658b74 to 2：新rs增加一个副本到2，合计4个副本。

4）Scaled down replica set nginx-66b6c48dd5 to 1：旧rs减少一个副本到1，合计3个副本。

5）Scaled up replica set nginx-559d658b74 to 3：新rs增加一个副本到3，合计4个副本。

6）Scaled down replica set nginx-66b6c48dd5 to 0：旧rs减少一个副本到0，合计3个副本。

接着来看回滚操作，我们先看如何查询更新历史：

```
# kubectl rollout history deployments/nginx
deployment.apps/nginx
REVISION    CHANGE-CAUSE
1           <none>
2           <none>
```

这里可以看到一个细节，CHANGE-CAUSE是空的，这个字段其实是从kubernetes.io/change-cause注解中提取的，我们加一个注解试试：

```
kubectl annotate deployment/nginx kubernetes.io/change-cause="image updated to 1.16.1"
```

然后查询一次：

```
# kubectl rollout history deployments/nginx
deployment.apps/nginx
REVISION    CHANGE-CAUSE
1           <none>
2           image updated to 1.16.1
```

第一个版本的注解似乎没有地方可以加？这里大概可以猜到如果要支持存储多个版本的CHANGE-CAUSE信息，这个注解应该是用于ReplicaSet中的，所以我们尝试这样补充第一个版本的注解：

```
kubectl annotate rs/nginx-66b6c48dd5 kubernetes.io/change-cause="nginx deployment created"
```

然后查询一次：

```
# kubectl rollout history deployments/nginx
```

```
deployment.apps/nginx
REVISION    CHANGE-CAUSE
1           nginx deployment created
2           image updated to 1.16.1
```

好了,现在每个版本都有注释信息了,这样我们进行回滚动作时就更加清楚需要指定哪个版本了。

我们通过设置一个不存在的镜像版本来模拟更新失败的场景:

```
# kubectl set image deployment/nginx nginx=nginx:1.161
deployment.apps/nginx image updated
# kubectl get rs --selector=app=nginx
NAME                DESIRED   CURRENT   READY   AGE
nginx-559d658b74    3         3         3       168m
nginx-66b6c48dd5    0         0         0       3h13m
nginx-66bc5d6c8     1         1         0       6s
# kubectl get pod --selector=app=nginx
NAME                        READY   STATUS             RESTARTS   AGE
nginx-559d658b74-l4bq7      1/1     Running            0          170m
nginx-559d658b74-qhh8m      1/1     Running            0          170m
nginx-559d658b74-vbtl5      1/1     Running            0          170m
nginx-66bc5d6c8-tl848       0/1     ImagePullBackOff   0          2m2s
```

给当前版本设置一个注解:

```
# kubectl annotate deployment/nginx kubernetes.io/change-cause="image updated to 1.161"
deployment.apps/nginx annotated
# kubectl rollout history deployments/nginx
deployment.apps/nginx
REVISION    CHANGE-CAUSE
1           nginx deployment created
2           image updated to 1.16.1
3           image updated to 1.161
```

然后尝试将版本回滚到2:

```
# kubectl rollout undo deployment/nginx
deployment.apps/nginx rolled back
# kubectl rollout history deployments/nginx
deployment.apps/nginx
REVISION    CHANGE-CAUSE
1           nginx deployment created
3           image updated to 1.161
4           image updated to 1.16.1
```

可以看到这时版本2变成了最新的版本:4。我们可以通过这种方式来查看某个版本的详细配置:

```
# kubectl rollout history deployments/nginx --revision=1
deployment.apps/nginx with revision #1
Pod Template:
  Labels:     app=nginx
```

```
              pod-template-hash=66b6c48dd5
Annotations:  kubernetes.io/change-cause: nginx deployment created
Containers:
 nginx:
  Image:         nginx:1.14.2
  Port:          80/TCP
  Host Port:     0/TCP
  Environment:   <none>
  Mounts:        <none>
 Volumes:        <none>
```

另外，回滚也可以指定具体的版本：

```
# kubectl rollout undo deployment/nginx --to-revision=1
deployment.apps/nginx rolled back
# kubectl rollout history deployments/nginx
deployment.apps/nginx
REVISION   CHANGE-CAUSE
3          image updated to 1.161
4          image updated to 1.16.1
5          nginx deployment created
```

8.1.3　Deployment 的其他特性

最后我们了解Deployment类型spec的全部属性：

- **minReadySeconds**：默认值为0，表示一个Pod 就绪之后多长时间可以提供服务。换句话说，配置成1就是Pod就绪之后1秒才对外提供服务。
- **paused**：挂起。
- **progressDeadlineSeconds**：默认值为600，表示处理一个Deployment任务的超时时间，比如10分钟到了还没有升级成功，则标记为failed（失败）状态。
- **replicas**：副本数。
- **revisionHistoryLimit**：默认是10，表示保留的历史版本数量。
- **selector**：标签选择器。
- **strategy**：表示Deployment更新Pod时的替换策略。
- **template**：Pod模板。

上面提到的strategy有两个属性，分别是type和rollingUpdate。Type的可选值是"Recreate"和"RollingUpdate"，默认为"RollingUpdate"。strategy.rollingUpdate有两个属性：

- **maxSurge**：表示滚动更新的时候最多可以比期望副本数多几个，数字或者百分比配置都行，比如1表示更新过程中最多同时新增1个副本，然后等一个旧副本删掉之后才能继续增加1个新副本，百分比计算的结果要向上取整。
- **maxUnavailable**：表示滚动更新的时候可以有多少副本不可用，同样是数字或者百分比配置，比如期望副本数是3，1表示最多删除副本到剩下2，然后要等新副本创建才能继续删除，百分比计算的结果要向下取整。

8.1.4 小结

本节我们的目的是知道Deployment的全部特性，进而为后面的源码分析做准备。在这个过程中没有赘述Deployment的基础特性，而是主要介绍"滚动更新"和"回滚"等主要功能，另外简单过一下Deployment的spec包含的全量配置项，从而心中有一个概念，知道Deployment的能力边界在哪里，从而后面看源码时更有针对性。

8.2 Deployment 源码分析

前一节介绍了Deployment的各种功能，本节继续从源码角度来看这些功能所对应的实现。

与client-go的源码分析部分一样，这部分的源码也是在2021年就完成了，并且整理了相关笔记，实时发布在了笔者的个人微信公众号"胡说云原生"上。为了尽量避免本书出版后大家看到的源码版本落后太多，笔者再次更新本地代码到最新版本，然后参考之前的笔记，重新整理一版"源码分析笔记"，这样大家看到本书时，大概率自己在工作中所使用的Kubernetes版本会和本章内容所基于的源码版本相差很小。

同时需要注意的是，阅读Deployment源码需要有一定的自定义控制器工作原理基础，里面涉及Informer工作机制、WorkQueue（延时/限速工作队列）、ResourceEventHandler等逻辑，也就是需要先完成第5章的学习。

本节内容的源码版本同样和client-go保持一致，也就是2022年2月3日23时所获取的最新版本的代码，分支是master。

8.2.1 逻辑入口：startDeploymentController

先进入Kubernetes项目的cmd/kube-controller-manager/app包，我们在apps.go中可以看到startDeploymentController()函数，这个函数也就是DeploymentController的初始化和启动入口：

```
func startDeploymentController(ctx context.Context, controllerContext
ControllerContext) (controller.Interface, bool, error) {
    dc, err := deployment.NewDeploymentController(
        controllerContext.InformerFactory.Apps().V1().Deployments(),
        controllerContext.InformerFactory.Apps().V1().ReplicaSets(),
        controllerContext.InformerFactory.Core().V1().Pods(),
        controllerContext.ClientBuilder.ClientOrDie("deployment-controller"),
    )
    if err != nil {
        return nil, true, fmt.Errorf("error creating Deployment controller: %v",
err)
    }
    go dc.Run(ctx, int(controllerContext.ComponentConfig.DeploymentController.
ConcurrentDeploymentSyncs))
    return nil, true, nil
}
```

在startDeploymentController()函数中先通过NewDeploymentController()方法初始化一个DeploymentController实例，这里的参数是DeploymentInformer、ReplicaSetInformer、PodInformer和ClientSet，因而DeploymentController也就具备了获取Deployment、ReplicaSet、Pod三类资源变更事件以及CURD apiserver操作各种资源的能力。接着这个函数中又调用了DeploymentController的Run()方法来启动DeploymentController，这里的参数ConcurrentDeploymentSyncs默认值是5，也就是默认情况下并发调谐的Deployment数量是5个。

8.2.2 DeploymentController 对象初始化

我们继续查看DeploymentController对象的定义和初始化过程，先看一下DeploymentController类型的定义。进入pkg/controller/deployment包，打开deployment_controller.go文件，可以看到DeploymentController的定义如下：

```go
type DeploymentController struct {
    // ReplicaSet操控器
    rsControl        controller.RSControlInterface
    client           clientset.Interface
    eventRecorder    record.EventRecorder

    syncHandler func(dKey string) error
    // 测试用
    enqueueDeployment func(deployment *apps.Deployment)

    // 用来从cache中get/list Deployment
    dLister appslisters.DeploymentLister
    // 用来从cache中get/list ReplicaSet
    rsLister appslisters.ReplicaSetLister
    // 用来从cache中get/list Pod
    podLister corelisters.PodLister

    dListerSynced cache.InformerSynced
    rsListerSynced cache.InformerSynced
    podListerSynced cache.InformerSynced

    // 工作队列，限速队列实现
    queue workqueue.RateLimitingInterface
}
```

DeploymentController初始化逻辑也在这个源文件中，我们继续来看NewDeploymentController函数的实现：

```go
func NewDeploymentController(dInformer appsinformers.DeploymentInformer,
rsInformer appsinformers.ReplicaSetInformer, podInformer coreinformers.PodInformer,
client clientset.Interface) (*DeploymentController, error) {
    // Event相关逻辑
    eventBroadcaster := record.NewBroadcaster()
    eventBroadcaster.StartStructuredLogging(0)
    eventBroadcaster.StartRecordingToSink(&v1core.EventSinkImpl{Interface:
client.CoreV1().Events("")})

    // ...
```

```go
    // 初始化一个DeploymentController对象实例
    dc := &DeploymentController{
        client:        client,
        eventRecorder: eventBroadcaster.NewRecorder(scheme.Scheme,
v1.EventSource{Component: "deployment-controller"}),
        queue:         workqueue.NewNamedRateLimitingQueue(workqueue.
DefaultControllerRateLimiter(), "deployment"),
    }
    // 主要是ClientSet
    dc.rsControl = controller.RealRSControl{
        KubeClient: client,
        Recorder:   dc.eventRecorder,
    }
    // ResourceEventHandler配置，后面会分析到
    dInformer.Informer().AddEventHandler(cache.ResourceEventHandlerFuncs{
        AddFunc:    dc.addDeployment,
        UpdateFunc: dc.updateDeployment,
        DeleteFunc: dc.deleteDeployment,
    })
    rsInformer.Informer().AddEventHandler(cache.ResourceEventHandlerFuncs{
        AddFunc:    dc.addReplicaSet,
        UpdateFunc: dc.updateReplicaSet,
        DeleteFunc: dc.deleteReplicaSet,
    })
    podInformer.Informer().AddEventHandler(cache.ResourceEventHandlerFuncs{
        DeleteFunc: dc.deletePod,
    })

    // 这里有主要逻辑，后面会继续分析到
    dc.syncHandler = dc.syncDeployment
    dc.enqueueDeployment = dc.enqueue
    // 各种lister
    dc.dLister = dInformer.Lister()
    dc.rsLister = rsInformer.Lister()
    dc.podLister = podInformer.Lister()
    dc.dListerSynced = dInformer.Informer().HasSynced
    dc.rsListerSynced = rsInformer.Informer().HasSynced
    dc.podListerSynced = podInformer.Informer().HasSynced
    return dc, nil
}
```

这里提到了ResourceEventHandler和syncDeployment相关的逻辑比较重要，在8.2.3节和8.2.4节继续来看这两个知识点。

8.2.3　ResourceEventHandler 逻辑

前文提到的ResourceEventHandler，具体来看回调函数有如下几个：

- addDeployment。
- updateDeployment。

- deleteDeployment。
- addReplicaSet。
- updateReplicaSet。
- deleteReplicaSet。
- deletePod。

本小节就来看一下这些回调函数的实现逻辑。

1. Deployment变更事件相关函数

还是看deployment_controller.go源文件，我们可以找到addDeployment、updateDeployment和deleteDeployment三个方法的代码：

```
func (dc *DeploymentController) addDeployment(obj interface{}) {
    d := obj.(*apps.Deployment)
    // ...
    // 新增 Deployment 时直接入队（enqueue）
    dc.enqueueDeployment(d)
}

func (dc *DeploymentController) updateDeployment(old, cur interface{}) {
    oldD := old.(*apps.Deployment)
    curD := cur.(*apps.Deployment)
    // ...
    // old Deployment 只是用来打印一个日志, cur Deployment enqueue
    dc.enqueueDeployment(curD)
}

func (dc *DeploymentController) deleteDeployment(obj interface{}) {
    d, ok := obj.(*apps.Deployment)
    if !ok { // 处理DeletedFinalStateUnknown场景
        // ...
    }
    // ...
    // 入队
    dc.enqueueDeployment(d)
}
```

2. ReplicaSet变更事件相关函数：Added

同样在deployment_controller.go源文件中可以使用Added函数：

```
func (dc *DeploymentController) addReplicaSet(obj interface{}) {
    rs := obj.(*apps.ReplicaSet)
    // 如果准备删除了，在重启的过程中就会收到Added事件，这时直接调用删除操作
    if rs.DeletionTimestamp != nil {
        dc.deleteReplicaSet(rs)
        return
    }

    // 查询对应的Deployment
    if controllerRef := metav1.GetControllerOf(rs); controllerRef != nil {
        d := dc.resolveControllerRef(rs.Namespace, controllerRef)
```

```go
        if d == nil {
            return
        }
        klog.V(4).InfoS("ReplicaSet added", "replicaSet", klog.KObj(rs))
        dc.enqueueDeployment(d)
        return
    }
    // 如果是一个孤儿ReplicaSet，则看一下是否能找到一个Deployment来领养
    ds := dc.getDeploymentsForReplicaSet(rs)
    if len(ds) == 0 {
        return
    }
    // ...
    // 一般只有一个Deployment，但是也不能排除多个的情况，所以这里用的是ds 列表，循环enqueue
    for _, d := range ds {
        dc.enqueueDeployment(d)
    }
}
```

3. ReplicaSet变更事件相关函数：Updated

```go
func (dc *DeploymentController) updateReplicaSet(old, cur interface{}) {
    curRS := cur.(*apps.ReplicaSet)
    oldRS := old.(*apps.ReplicaSet)
    if curRS.ResourceVersion == oldRS.ResourceVersion {
        // Resync的时候RV相同，不做处理
        return
    }

    curControllerRef := metav1.GetControllerOf(curRS)
    oldControllerRef := metav1.GetControllerOf(oldRS)
    controllerRefChanged := !reflect.DeepEqual(curControllerRef, oldControllerRef)
    if controllerRefChanged && oldControllerRef != nil {
        // 如果rs的ref变更了，就需要通知旧的ref对应的Deployment
        if d := dc.resolveControllerRef(oldRS.Namespace, oldControllerRef); d != nil {
            dc.enqueueDeployment(d)
        }
    }

    if curControllerRef != nil {
        d := dc.resolveControllerRef(curRS.Namespace, curControllerRef)
        if d == nil {
            return
        }
        // ...
        // 当前rs对应dp入队
        dc.enqueueDeployment(d)
        return
    }

    // 孤儿rs的场景，与Added时处理逻辑一样
    labelChanged := !reflect.DeepEqual(curRS.Labels, oldRS.Labels)
```

```go
    if labelChanged || controllerRefChanged {
        ds := dc.getDeploymentsForReplicaSet(curRS)
        if len(ds) == 0 {
            return
        }
        // ...
        for _, d := range ds {
            dc.enqueueDeployment(d)
        }
    }
}
```

4. ReplicaSet变更事件相关函数：Deleted

```go
func (dc *DeploymentController) deleteReplicaSet(obj interface{}) {
    rs, ok := obj.(*apps.ReplicaSet)
    // 删除场景需要处理的DeletedFinalStateUnknown场景
    if !ok {
        // ...
    }
    // 孤儿rs被删除时没有Deployment需要关心
    controllerRef := metav1.GetControllerOf(rs)
    if controllerRef == nil {
        return
    }
    d := dc.resolveControllerRef(rs.Namespace, controllerRef)
    if d == nil {
        return
    }
    klog.V(4).InfoS("ReplicaSet deleted", "replicaSet", klog.KObj(rs))
    // 入队
    dc.enqueueDeployment(d)
}
```

8.2.4 DeploymentController 的启动过程

通过前几个小节的学习，我们已经知道了哪些Event会向WorkQueue中添加元素，接着看一下这些元素是怎么被消费的。

1. Run()函数

Run()方法本身的逻辑很简洁，根据给定的并发数（默认5并发）启动dc.worker。相关源码还是在deployment_controller.go源文件中：

```go
func (dc *DeploymentController) Run(ctx context.Context, workers int) {
    defer utilruntime.HandleCrash()
    defer dc.queue.ShutDown()
    // ...

    if !cache.WaitForNamedCacheSync("deployment", ctx.Done(), dc.dListerSynced,
dc.rsListerSynced, dc.podListerSynced) {
```

```
        return
    }
    for i := 0; i < workers; i++ {
        go wait.UntilWithContext(ctx, dc.worker, time.Second)
    }
    <-ctx.Done()
}
```

继续来看这里的worker的逻辑：

```
func (dc *DeploymentController) worker(ctx context.Context) {
    for dc.processNextWorkItem(ctx) {
    }
}
func (dc *DeploymentController) processNextWorkItem(ctx context.Context) bool {
    // 从WorkQueue中获取一个元素
    key, quit := dc.queue.Get()
    if quit {
        return false
    }
    defer dc.queue.Done(key)
    // 主要逻辑
    err := dc.syncHandler(ctx, key.(string))
    dc.handleErr(err, key)

    return true
}
```

这里从WorkQueue中拿到一个key（键）之后，通过调用syncHandler()方法来处理。前面我们强调过syncDeployment是一个重点，还记得那一行代码吗？如下所示：

```
dc.syncHandler = dc.syncDeployment
```

也就是说，这里的dc.syncHandler(ctx, key.(string))调用的本质是dc.syncDeployment()，所以接着继续dc.syncDeployment的实现。

2. syncDeployment()方法

syncDeployment()方法完成的事情是获取从WorkQueue中出队的key，根据这个key来sync（同步）对应的Deployment。下面来看具体的代码逻辑：

```
func (dc *DeploymentController) syncDeployment(ctx context.Context, key string) error {
    // 从key中分割出namespace和name
    namespace, name, err := cache.SplitMetaNamespaceKey(key)
    if err != nil {
        klog.ErrorS(err, "Failed to split meta namespace cache key", "cacheKey", key)
        return err
    }
    // ...
```

```go
        // 根据namespace和name从cache中检索对应的Deployment对象
        deployment, err := dc.dLister.Deployments(namespace).Get(name)
        if errors.IsNotFound(err) {
            klog.V(2).InfoS("Deployment has been deleted", "deployment", klog.KRef(namespace, name))
            return nil
        }
        if err != nil {
            return err
        }

        // 为了不改动这个cache,这是一个ThreadSafeStore
        d := deployment.DeepCopy()

        // 空LabelSelector会匹配到所有Pod,发出一个警告事件(Warning Event),更新d.Status.ObservedGeneration然后返回
        everything := metav1.LabelSelector{}
        if reflect.DeepEqual(d.Spec.Selector, &everything) {
            dc.eventRecorder.Eventf(d, v1.EventTypeWarning, "SelectingAll", "This deployment is selecting all pods. A non-empty selector is required.")
            if d.Status.ObservedGeneration < d.Generation {
                d.Status.ObservedGeneration = d.Generation
                dc.client.AppsV1().Deployments(d.Namespace).UpdateStatus(ctx, d, metav1.UpdateOptions{})
            }
            return nil
        }

        // 获取当前Deployment拥有的所有ReplicaSet,同时会更新这些ReplicaSet的ControllerRef
        rsList, err := dc.getReplicaSetsForDeployment(ctx, d)
        if err != nil {
            return err
        }
        // 这个map是map[types.UID][]*v1.Pod类型,key是rs的UID,value是对应rs管理的所有Pod列表
        podMap, err := dc.getPodMapForDeployment(d, rsList)
        if err != nil {
            return err
        }
        // 已经标记要删除了,这时只更新状态
        if d.DeletionTimestamp != nil {
            return dc.syncStatusOnly(ctx, d, rsList)
        }
        // 根据d.Spec.Pause配置看是否更新Deployment的conditions
        if err = dc.checkPausedConditions(ctx, d); err != nil {
            return err
        }

        if d.Spec.Paused {
            // Pause或scale时的调谐逻辑
            return dc.sync(ctx, d, rsList)
        }
```

```go
    // 应该过期了，旧版本的deprecated.deployment.rollback.to注解回滚逻辑
    if getRollbackTo(d) != nil {
        // 回滚到旧版本的逻辑
        return dc.rollback(ctx, d, rsList)
    }
    // 如果是scale
    scalingEvent, err := dc.isScalingEvent(ctx, d, rsList)
    if err != nil {
        return err
    }
    // Pause或scale时的调谐逻辑
    if scalingEvent {
        return dc.sync(ctx, d, rsList)
    }
    switch d.Spec.Strategy.Type {
    // 重建策略
    case apps.RecreateDeploymentStrategyType:
        return dc.rolloutRecreate(ctx, d, rsList, podMap)
    // 滚动更新策略
    case apps.RollingUpdateDeploymentStrategyType:
        return dc.rolloutRolling(ctx, d, rsList)
    }
    return fmt.Errorf("unexpected deployment strategy type: %s", d.Spec.Strategy.Type)
}
```

看完syncDeployment()方法之后，Deployment控制器的逻辑就算过完一遍了。当然，这个方法内部涉及的一些小方法的调用这里只是简单介绍其功能，并没有深究所有实现细节，不过这些小方法的逻辑都不难，这里不再赘述了。

8.3 本章小结

本章先分析了Deployment的各种功能特性，然后进一步详细分析了Deployment控制器的源码实现。通过对Deployment控制器源码的学习，我们可以知道Kubernetes项目本身是如何实现控制器逻辑的，进而指导自己的Operator项目开发。

当然，一个Deployment源码肯定无法覆盖Operator开发的所有技巧，建议大家花些时间多阅读一些优秀的开源Operator项目源码。

Kubernetes Operator 开发进阶

第三篇

工 具

第 9 章 使用 Kustomize 管理配置

我们使用Kubebuilder生成的Operator脚手架中有很多资源配置文件，大家应该已经注意到这些资源配置文件都是使用Kustomize来管理的，所以为了完全看懂或者修改这些配置文件，有必要好好学习一下Kustomize工具的使用。

Kustomize的代码库GitHub地址是https://github.com/kubernetes-sigs/kustomize。Kustomize是Kubernetes原生配置管理工具，实现了类似sed的给资源配置文件"打补丁"的能力。Kubernetes 1.14.0版本已经集成到了kubectl命令中，所以也可以通过kubectl的子命令来使用Kustomize，如图9-1所示。

图 9-1 Kubectl 与 Kustomize

9.1 Kustomize 的基本概念

在学习Kustomize的用法之前，先来认识几个关键术语：

- kustomization：这个词指代的是一个kustomization.yaml文件，或者更广义地理解为一个包含kustomization.yaml文件的目录以及这个kustomization.yaml文件中引用的所有其他文件。
- base：指的是被其他kustomization引用的一个kustomization。换言之，任何一个kustomization a被另一个kustomization b引用时，a是b的base，这时如果新增一个kustomization c来引用b，那么b也就是c的base。也就是说，base是一个相对的概念，而不是某种属性标识。

- overlay：与base相对应，依赖另一个kustomization的kustomization被称为overlay。也就是说，如果kustomization b引用了kustomization a，那么b是a的overlay，a是b的base。

我们再通过一个小例子来直观认识kustomization这个概念。比如现在创建一个目录myapp，然后在其中创建一个base子目录，base目录中又有如下几个文件：

- kustomization.yaml。
- nginx-deployment.yaml。
- nginx-service.yaml。

其中kustomization.yaml文件内容如下：

```
resources:
- nginx-deployment.yaml
- nginx-service.yaml
```

这就是一个很基础的kustomization，包含一个kustomization.yaml文件，并且在kustomization.yaml中引用了两个资源文件。

然后我们继续在base同级目录下创建一个overlays目录，并且在overlays目录中放置dev和prod两个子目录，其目录结构大致如下：

```
/myapp
├── base
│   ├── kustomization.yaml
│   ├── nginx-deployment.yaml
│   └── nginx-service.yaml
└── overlays
    ├── dev
    └── prod
```

这时读者可能已经猜到将在dev和prod目录下分别创建kustomization.yaml，并且以base kustomization为基础，补充"开发环境"和"生产环境"的差异化配置，形成两个overlay kustomization。

接着在dev和prod的kustomization.yaml中都放置这样的内容：

```
bases:
- ../../base
patchs:
- replica.yaml
```

这里我们指定bases是前面的base kustomization，然后指定了一个patchs replica.yaml。接着dev和prod目录中可以放置不同的replica.yaml，其中唯一的区别就是副本数不同，开发环境指定1个Pod副本，而生产环境指定3个Pod副本。我们在这里不详细地展示replica.yaml和其他几个资源配置YAML文件的具体内容，现在的目的是希望大家对于Kustomize的作用和能力有一个初步的认识，后面会有详细的例子讲解Kustomize的各种使用细节。最后查看最新的目录结构：

```
/myapp
├── base
│   ├── kustomization.yaml
```

```
        │     ├── nginx-deployment.yaml
        │     └── nginx-service.yaml
        └── overlays
            ├── dev
            │     ├── kustomization.yaml
            │     └── replica.yaml
            └── prod
                  ├── kustomization.yaml
                  └── replica.yaml
```

简单来说，Kustomize的作用就是提供了一种机制让我们能够管理同一套应用资源配置文件在不同环境的差异化配置问题，将共性部分抽象成base，然后通过overlay的方式针对不同使用场景打补丁，从而增量得到多种"场景化"配置。

9.2　Kustomize 的安装

Kustomize提供了Linux/Darwin系统* amd64/arm64架构的二进制可执行文件（Windows等也支持，不过不建议在Windows上使用这些工具）。非Darwin - arm64的用户可以直接执行下面的命令下载并安装：

```
curl -s \
"https://raw.githubusercontent.com/kubernetes-sigs/\
kustomize/master/hack/install_kustomize.sh" | bash
```

笔者用的是arm64的MacBook，这个脚本暂时不支持，所以需要到Kustomize项目的release页面https://github.com/kubernetes-sigs/kustomize/releases去下载对应的压缩包，如图9-2所示。

图 9-2　下载压缩包

根据自己的开发环境下载对应的压缩包：

```
tar -xzvf kustomize_v4.4.1_darwin_arm64.tar.gz
sudo mv kustomize /usr/local/bin
kustomize version
```

9.3 使用 Kustomize 生成资源

我们在Kubernetes中通常使用ConfigMap和Secret来分别存储配置文件和敏感配置信息等，这些内容往往是在Kubernetes集群之外的。比如通过Secret来存储数据库连接信息，这些信息也许记录在特定机器的环境变量中，也许保存在某台机器的一个TXT文本文件中，总之这些信息和Kubernetes集群本身没有关联。但是我们的应用以Pod的方式运行在一个Kubernetes集群之内时，就需要使用ConfigMap或者Secret资源对象来获取各种配置，怎么快速创建ConfigMap和Secret呢？

9.3.1 ConfigMap 生成器

我们可以通过configMapGenerator来自动管理ConfigMap资源文件的创建和引用等。下面分别通过几个实例来演示相关的用法。

1. 从配置文件生成ConfigMap

```
mkdir kustomize-examples
cd kustomize-examples
cat <<EOF >config.txt
key=value
EOF

cat <<EOF >./kustomization.yaml
configMapGenerator:
- name: app-config
  files:
  - config.txt
EOF
```

然后有两种方式来构建ConfigMap：

```
# kustomize build .
apiVersion: v1
data:
  config.txt: |
    key=value
kind: ConfigMap
metadata:
  name: app-config-gc6cm9fg4c
# kubectl kustomize .
apiVersion: v1
```

```
data:
  config.txt: |
    key=value
kind: ConfigMap
metadata:
  name: app-config-gc6cm9fg4c
```

可以看到kustomize build <kustomization_directory>和kubectl kustomize <kustomization_directory>都可以输出想要的资源配置。本节后面的示例都以kustomize build命令为例。

除了可以从文本文件生成ConfigMap之外，也可以使用环境变量中的配置内容。

2. 通过环境变量创建ConfigMap

```
cat <<EOF >golang_env.txt
GOVERSION=go1.17.4
GOARCH
EOF

cat <<EOF >./kustomization.yaml
configMapGenerator:
- name: app-config
  envs:
  - golang_env.txt
EOF
```

接着构建ConfigMap：

```
# GOARCH=arm64 kustomize build .
apiVersion: v1
data:
  GOARCH: arm64
  GOVERSION: go1.17.4
kind: ConfigMap
metadata:
  name: app-config-g4tdtggbd7
```

可以看到在golang_env.txt中存放的key=value格式的GOVERSION配置以及环境变量中的GOARCH都包含在这个ConfigMap中了。

3. 通过键值对字面值直接创建ConfigMap

```
cat <<EOF >./kustomization.yaml
configMapGenerator:
- name: app-config
  literals:
  - Hello=World
EOF
```

接着构建ConfigMap看一下效果：

```
# kustomize build .
apiVersion: v1
data:
  Hello: World
```

```
kind: ConfigMap
metadata:
  name: app-config-7b4b2hf646
```

4. 使用ConfigMap

大家可以注意到一个问题，就是通过Kustomize生成的ConfigMap的名称默认带了一串后缀，那么在Deployment中引用这个ConfigMap的时候，怎么预知这个名字呢？其实 Kustomize会自动在Deployment配置中替换这个字段，我们看一下具体的例子。

配置文件config.txt的内容如下：

```
cat <<EOF >config.txt
key=value
EOF
```

Deployment配置文件nginx-deployment.yaml的内容如下：

```
cat <<EOF >nginx-deployment.yaml
apiVersion: apps/v1
kind: Deployment
metadata:
  name: nginx
  labels:
    app: nginx
spec:
  selector:
    matchLabels:
      app: nginx
  template:
    metadata:
      labels:
        app: nginx
    spec:
      containers:
      - name: nginx
        image: nginx:1.16
        volumeMounts:
        - name: config
          mountPath: /config
      volumes:
      - name: config
        configMap:
          name: app-config
EOF
```

kustomization.yaml文件的内容如下：

```
cat <<EOF >./kustomization.yaml
resources:
- nginx-deployment.yaml
configMapGenerator:
- name: app-config
```

```
    files:
    - config.txt
EOF
```

最后看构建出来的资源是什么样的：

```
# kustomize build .
apiVersion: v1
data:
  config.txt: |
    key=value
kind: ConfigMap
metadata:
  name: app-config-gc6cm9fg4c
---
apiVersion: apps/v1
kind: Deployment
metadata:
  labels:
    app: nginx
  name: nginx
spec:
  selector:
    matchLabels:
      app: nginx
  template:
    metadata:
      labels:
        app: nginx
    spec:
      containers:
      - image: nginx:1.16
        name: nginx
        volumeMounts:
        - mountPath: /config
          name: config
      volumes:
      - configMap:
          name: app-config-gc6cm9fg4c
        name: config
```

可以看到生成的ConfigMap名为app-config-gc6cm9fg4c，同时Deployment部分的.spec.template.spec.volumes[0].configMap.name也对应配置成了app-config-gc6cm9fg4c。

9.3.2　Secret生成器

与ConfigMap类似，有好几种方式来生成Secret资源配置。

1. 通过配置文件生成Secret

```
cat <<EOF >./password.txt
username=daniel
password=leinad
```

```
EOF
cat <<EOF >./kustomization.yaml
secretGenerator:
- name: app-secret
  files:
  - password.txt
EOF
```

查看创建的Secret：

```
# kustomize build .
apiVersion: v1
data:
  password.txt: dXNlcm5hbWU9ZGFuaWVsCnBhc3N3b3JkPWxlaW5hZAo=
kind: Secret
metadata:
  name: app-secret-25d45kc475
type: Opaque
```

可以看到一个叫作app-secret-25d45kc475的Secret类型资源配置被输出了，其中的内容是一串base64编码后的字符串，我们可以解码这串字符看一下内容是否符合预期：

```
# echo "dXNlcm5hbWU9ZGFuaWVsCnBhc3N3b3JkPWxlaW5hZAo=" \
| base64 -d
username=daniel
password=leinad
```

2. 通过键值对字面值创建Secret

```
cat <<EOF >./kustomization.yaml
secretGenerator:
- name: app-secret
  literals:
  - username=daniel
  - password=leinad
EOF
```

查看创建的Secret：

```
kustomize build .
apiVersion: v1
data:
  password: bGVpbmFk
  username: ZGFuaWVs
kind: Secret
metadata:
  name: app-secret-22696gfm7h
type: Opaque
```

同样可以用base64解码看一下内容是否正确：

```
# echo "ZGFuaWVs" | base64 -d
daniel%
```

```
# echo "bGVpbmFk" | base64 -d
leinad%
```

也许大家会注意到这里多出了一个%符号,这是自动加在没有换行符的字符串结尾的,然后自动强制换行,能够让下一个输出从新的一行开始,让终端不至于乱糟糟的。

3. 使用Secret

与ConfigMap的用法类似,同样可以在Deployment中使用带后缀的Secret。

先准备Secret配置:

```
cat <<EOF >./password.txt
username=daniel
password=leinad
EOF
```

然后准备一个Deployment:

```
cat <<EOF >nginx-deployment.yaml
apiVersion: apps/v1
kind: Deployment
metadata:
  name: nginx
  labels:
    app: nginx
spec:
  selector:
    matchLabels:
      app: nginx
  template:
    metadata:
      labels:
        app: nginx
    spec:
      containers:
      - name: nginx
        image: nginx:1.16
        volumeMounts:
        - name: password
          mountPath: /secrets
      volumes:
      - name: password
        secret:
          secretName: app-secret
EOF
```

最后需要一个kustomization.yaml:

```
cat <<EOF >./kustomization.yaml
resources:
- nginx-deployment.yaml
secretGenerator:
- name: app-secret
```

```
      files:
      - password.txt
EOF
```

查看一下成果：

```
# kustomize build .
apiVersion: v1
data:
  password.txt: dXNlcm5hbWU9ZGFuaWVsCnBhc3N3b3JkPWxlaW5hZAo=
kind: Secret
metadata:
  name: app-secret-25d45kc475
type: Opaque
---
apiVersion: apps/v1
kind: Deployment
metadata:
  labels:
    app: nginx
  name: nginx
spec:
  selector:
    matchLabels:
      app: nginx
  template:
    metadata:
      labels:
        app: nginx
    spec:
      containers:
      - image: nginx:1.16
        name: nginx
        volumeMounts:
        - mountPath: /secrets
          name: password
      volumes:
      - name: password
        secret:
          secretName: app-secret-25d45kc475
```

可以看到生成的Secret名为app-secret-25d45kc475，同时Deployment部分的.spec.template.spec.volumes[0].secret.secretName也对应配置成app-secret-25d45kc475，和ConfigMap的处理方式基本一致。

9.3.3　使用generatorOptions改变默认行为

前面使用ConfigMap和Secret生成器时，大家已经注意到最终生成的资源配置名字上会有一串随机字符串，这个行为的意义是为了保证不同配置内容生成的资源名字会不一样，减少误用的概率。如果读者不喜欢或者不需要这种默认行为又怎么办呢？Kustomize其实提供了开关，可以通过generatorOptions来改变这种行为。查看下面的这个例子：

```
cat <<EOF >./kustomization.yaml
configMapGenerator:
- name: app-config
  literals:
  - Hello=World
generatorOptions:
  disableNameSuffixHash: true
  labels:
    type: generated
  annotations:
    note: generated
EOF
```

看一下构建结果：

```
# kustomize build .
apiVersion: v1
data:
  Hello: World
kind: ConfigMap
metadata:
  annotations:
    note: generated
  labels:
    type: generated
  name: app-config
```

可以看到这次得到的资源名字变成了没有后缀的app-config，同时这里演示了generatorOptions可以给资源统一添加labels和annotations。

9.4　使用 Kustomize 管理公共配置项

我们经常需要在不同的资源配置文件中配置相同的字段，比如：

- 给所有的资源配置相同的namespace。
- 给多个资源的name字段加上相同的前缀或者后缀。
- 给多个资源配置相同的labels或annotations。

......

当我们需要修改某一项配置，比如临时改变主意，想要将资源部署在另一个namespace中，这时需要编辑几乎所有的资源配置文件。你猜得没错，Kustomize也可以解决这类问题，统一管理这种公共配置项。

我们先准备一个普通的Deployment模板：

```
cat <<EOF >./nginx-deployment.yaml
apiVersion: apps/v1
kind: Deployment
metadata:
```

```
  name: nginx
  labels:
    app: nginx
spec:
  selector:
    matchLabels:
      app: nginx
  template:
    metadata:
      labels:
        app: nginx
    spec:
      containers:
      - name: nginx
        image: nginx:1.16
EOF
```

这个配置平淡无奇,我们给它加一些配置项:

```
cat <<EOF >./kustomization.yaml
namespace: user-daniel
namePrefix: app-
nameSuffix: -v1
commonLabels:
  version: v1
commonAnnotations:
  owner: daniel
resources:
- nginx-deployment.yaml
EOF
```

查看一下构建结果:

```
kustomize build .
apiVersion: apps/v1
kind: Deployment
metadata:
  annotations:
    owner: daniel
  labels:
    app: nginx
    version: v1
  name: app-nginx-v1
  namespace: user-daniel
spec:
  selector:
    matchLabels:
      app: nginx
      version: v1
  template:
    metadata:
      annotations:
        owner: daniel
```

```
      labels:
        app: nginx
        version: v1
    spec:
      containers:
      - image: nginx:1.16
        name: nginx
```

可以看到定义的namespace、name前后缀、label和annotation都生效了。使用这种方式就可以将多个资源的一些公共配置抽取出来，以便于管理。

9.5 使用 Kustomize 组合资源

通过Kustomize可以灵活组合多个资源或者给多个资源"打补丁"从而拓展配置。本节来学习一下相关的用法。

9.5.1 多个资源的组合

很多时候在Kubernetes上部署一个应用时需要用到多个资源类型的配置，比如Deployment和Service，它们往往通过不同的文件来保存，比如nginx-deployment.yaml和nginx-service.yaml。我们看一下如何通过kustomize来组合这两种配置。

先准备一个Deployment配置：

```
cat <<EOF > nginx-deployment.yaml
apiVersion: apps/v1
kind: Deployment
metadata:
  name: nginx
spec:
  selector:
    matchLabels:
      app: nginx
  replicas: 3
  template:
    metadata:
      labels:
        app: nginx
    spec:
      containers:
      - name: nginx
        image: nginx:1.16
        ports:
        - containerPort: 80
```

然后准备一个Service配置：

```
cat <<EOF > nginx-service.yaml
apiVersion: v1
```

```
kind: Service
metadata:
  name: nginx
  labels:
    app: nginx
spec:
  ports:
  - port: 80
    protocol: TCP
  selector:
    app: nginx
EOF
```

最后编写 kustomization.yaml：

```
cat <<EOF >./kustomization.yaml
resources:
- nginx-deployment.yaml
- nginx-service.yaml
EOF
```

看一下构建的结果：

```
# kustomize build .
apiVersion: v1
kind: Service
metadata:
  labels:
    app: nginx
  name: nginx
spec:
  ports:
  - port: 80
    protocol: TCP
  selector:
    app: nginx
---
apiVersion: apps/v1
kind: Deployment
metadata:
  labels:
    app: nginx
  name: nginx
spec:
  selector:
    matchLabels:
      app: nginx
  template:
    metadata:
      labels:
        app: nginx
    spec:
      containers:
```

```
        - image: nginx:1.16
          name: nginx
```

9.5.2 给资源配置打补丁

很多时候需要给同一个资源针对不同使用场景配置不同的配置项。比如同样一个nginx应用，可能在开发环境需要100MB的内存就足够了，但是在生产环境我们希望"大方"一些，提供1GB，这时如果分别使用两个配置文件来保存开发环境和生产环境的nginx配置，明显是不够优雅的。在理解Kustomize如何解决这类问题之前，我们先看一下可以针对一个资源做哪些"打补丁"的操作，进而通过给一个资源"打不同的补丁"来实现"多环境配置灵活管理"。

1. patchesStrategicMerge方式自定义配置

同样先准备一个普通的Deployment配置文件：

```
cat <<EOF > nginx-deployment.yaml
apiVersion: apps/v1
kind: Deployment
metadata:
  name: nginx
spec:
  selector:
    matchLabels:
      app: nginx
  replicas: 3
  template:
    metadata:
      labels:
        app: nginx
    spec:
      containers:
      - name: nginx
        image: nginx:1.16
        ports:
        - containerPort: 80
EOF
```

然后单独将内存配置放到一个新的文件中：

```
cat <<EOF > nginx-memory.yaml
apiVersion: apps/v1
kind: Deployment
metadata:
  name: nginx
spec:
  template:
    spec:
      containers:
      - name: nginx
        resources:
```

```
          limits:
            memory: 100Mi
EOF
```

接着编写kustomization.yaml：

```
cat <<EOF >./kustomization.yaml
resources:
- nginx-deployment.yaml
patchesStrategicMerge:
- nginx-memory.yaml
EOF
```

查看一下构建结果：

```
# kustomize build .
apiVersion: apps/v1
kind: Deployment
metadata:
  name: nginx
spec:
  replicas: 3
  selector:
    matchLabels:
      app: nginx
  template:
    metadata:
      labels:
        app: nginx
    spec:
      containers:
      - image: nginx:1.16
        name: nginx
        ports:
        - containerPort: 80
        resources:
          limits:
            memory: 100Mi
```

这种方式在kustomization.yaml中的patchesStrategicMerge部分列出的是补丁文件列表。需要注意的是，这些文件中描述的是同一个资源对象才行，一般实践是每个patch都实现一个明确的小功能，比如设置资源QoS是一个单独的补丁（patch），设置亲和性策略是一个单独的补丁，设置副本数又是一个单独的补丁，等等。

2. patchesJson6902方式自定义配置

同样以一个简单的Deployment配置为例，我们通过patchesJson6902的方式来patch这个Deployment的副本数。

准备一个Deployment：

```
cat <<EOF > nginx-deployment.yaml
apiVersion: apps/v1
```

```
kind: Deployment
metadata:
  name: nginx
spec:
  selector:
    matchLabels:
      app: nginx
  replicas: 3
  template:
    metadata:
      labels:
        app: nginx
    spec:
      containers:
      - name: nginx
        image: nginx:1.16
        ports:
        - containerPort: 80
EOF
```

然后给出一个patch配置文件：

```
cat <<EOF > patch.yaml
- op: replace
  path: /spec/replicas
  value: 1
EOF
```

接着编写kustomization.yaml文件：

```
cat <<EOF >./kustomization.yaml
resources:
- nginx-deployment.yaml

patchesJson6902:
- target:
    group: apps
    version: v1
    kind: Deployment
    name: nginx
  path: patch.yaml
EOF
```

最后看一下构建的效果：

```
# kustomize build .
apiVersion: apps/v1
kind: Deployment
metadata:
  name: nginx
spec:
  replicas: 1
  selector:
    matchLabels:
```

```
      app: nginx
  template:
    metadata:
      labels:
        app: nginx
    spec:
      containers:
      - image: nginx:1.16
        name: nginx
        ports:
        - containerPort: 80
```

结果符合预期，replicas字段更新了。这种方式需要注意的是，在kustomization.yaml中需要正确指定target，也就是group、version、kind、name等字段需要和patch的资源完全匹配才行。

3. 镜像的自定义

我们可以直接在kustomization.yaml中使用images配置来指定镜像。同样先创建一个Deployment用来测试：

```
cat <<EOF > nginx-deployment.yaml
apiVersion: apps/v1
kind: Deployment
metadata:
  name: nginx
spec:
  selector:
    matchLabels:
      app: nginx
  replicas: 3
  template:
    metadata:
      labels:
        app: nginx
    spec:
      containers:
      - name: nginx
        image: nginx:1.16
        ports:
        - containerPort: 80
EOF
```

然后在kustomization.yaml中通过images配置来指定一个新镜像：

```
cat <<EOF >./kustomization.yaml
resources:
- nginx-deployment.yaml
images:
- name: nginx
  newName: nginx
  newTag: 1.16.1
EOF
```

接着看一下构建的效果：

```
# kustomize build .
apiVersion: apps/v1
kind: Deployment
metadata:
  name: nginx
spec:
  replicas: 3
  selector:
    matchLabels:
      app: nginx
  template:
    metadata:
      labels:
        app: nginx
    spec:
      containers:
      - image: nginx:1.16.1
        name: nginx
        ports:
        - containerPort: 80
```

可以看到和预期的一样，image相关配置已经更新了。

4. 容器内使用其他资源对象的配置

还有一种场景是这样的，比如一个容器化应用启动时需要知道某个Service的名字，也许这个Service是该应用依赖的一个上游服务，所以拿到Service名字才能访问这个上游服务。在使用Kustomize之前，也许Service名字会通过硬编码的方式配置在YAML文件中。现在这个Service通过Kustomize来构建，它的名字也许会多出来一些前后缀，这时怎么动态获取这里的Service名字用于配置自己的Deployment呢？

查看如下的Deployment配置，注意其中的command部分相关配置：

```
cat <<'EOF' > nginx-deployment.yaml
apiVersion: apps/v1
kind: Deployment
metadata:
  name: nginx
spec:
  selector:
    matchLabels:
      app: nginx
  replicas: 3
  template:
    metadata:
      labels:
        app: nginx
    spec:
      containers:
      - name: nginx
```

```
      image: nginx:1.16
      command: ["start", "--host", "$(SERVICE_NAME)"]
      ports:
      - containerPort: 80
EOF
```

这里需要一个SERVICE_NAME，然后看一下Service的配置文件：

```
cat <<EOF > nginx-service.yaml
apiVersion: v1
kind: Service
metadata:
  name: nginx
  labels:
    app: nginx
spec:
  ports:
  - port: 80
    protocol: TCP
  selector:
    app: nginx
EOF
```

默认配置下Service的名字是nginx，然后查看kustomization.yaml：

```
cat <<EOF >./kustomization.yaml
namePrefix: dev-

resources:
- nginx-deployment.yaml
- nginx-service.yaml

vars:
- name: SERVICE_NAME
  objref:
    kind: Service
    name: nginx
    apiVersion: v1
EOF
```

这里给两个资源都加了一个dev-名字前缀，所以Service的名字就变成了dev-nginx。然后通过vars来定义SERVICE_NAME变量，该变量通过下面的objref内的几个配置项和上面的Service关联，最后看一下构建的结果：

```
# kustomize build .
apiVersion: v1
kind: Service
metadata:
  labels:
    app: nginx
  name: dev-nginx
spec:
  ports:
  - port: 80
```

```
      protocol: TCP
  selector:
    app: nginx
---
apiVersion: apps/v1
kind: Deployment
metadata:
  name: dev-nginx
spec:
  replicas: 3
  selector:
    matchLabels:
      app: nginx
  template:
    metadata:
      labels:
        app: nginx
    spec:
      containers:
      - command:
        - start
        - --host
        - dev-nginx
        image: nginx:1.16
        name: nginx
        ports:
        - containerPort: 80
```

可以看到最后command中用到的SERVICE_NAME变量被渲染成dev-nginx了，和预期一致。

9.6 Base 和 Overlay

前面已经介绍过Base和Overlay的概念，这里再补充一些信息。首先Base对Overlay的存在是无感的，Overlay引用的Base也不一定是一个本地目录，远程代码库的目录也可以，一个Overlay也可以有多个Base。我们通过一个具体的例子再来看一下Base和Overlay的用法。

首先准备一个Base目录，然后在Base目录内执行如下命令创建kustomization.yaml：

```
cat <<EOF > kustomization.yaml
resources:
- nginx-deployment.yaml
- nginx-service.yaml
EOF
```

接着创建这里引用的两个资源文件：

```
cat <<EOF > nginx-deployment.yaml
apiVersion: apps/v1
kind: Deployment
metadata:
```

```
      name: nginx
spec:
  selector:
    matchLabels:
      app: nginx
  replicas: 3
  template:
    metadata:
      labels:
        app: nginx
    spec:
      containers:
      - name: nginx
        image: nginx:1.16
EOF
```

nginx-service.yaml 的文件内容如下：

```
cat <<EOF > nginx-service.yaml
apiVersion: v1
kind: Service
metadata:
  name: nginx
  labels:
    app: nginx
spec:
  ports:
  - port: 80
    protocol: TCP
  selector:
    app: nginx
EOF
```

至此，Base就准备好了。然后可以创建两个Overlay来引用这个Base，并且打上不一样的name前缀来得到两套配置。

在Base同级目录下执行下面的命令：

```
mkdir dev
cat <<EOF > dev/kustomization.yaml
bases:
- ../base
namePrefix: dev-
EOF
mkdir prod
cat <<EOF > prod/kustomization.yaml
bases:
- ../base
namePrefix: prod-
EOF
```

最后查看一下构建的效果：

```
# kustomize build dev
apiVersion: v1
kind: Service
metadata:
  labels:
    app: nginx
  name: dev-nginx
spec:
  ports:
  - port: 80
    protocol: TCP
  selector:
    app: nginx
---
apiVersion: apps/v1
kind: Deployment
metadata:
  name: dev-nginx
spec:
  replicas: 3
  selector:
    matchLabels:
      app: nginx
  template:
    metadata:
      labels:
        app: nginx
    spec:
      containers:
      - image: nginx:1.16
        name: nginx
```

再查看prod的配置：

```
# kustomize build prod
apiVersion: v1
kind: Service
metadata:
  labels:
    app: nginx
  name: prod-nginx
spec:
  ports:
  - port: 80
    protocol: TCP
  selector:
    app: nginx
---
apiVersion: apps/v1
kind: Deployment
metadata:
  name: prod-nginx
```

```
spec:
  replicas: 3
  selector:
    matchLabels:
      app: nginx
  template:
    metadata:
      labels:
        app: nginx
    spec:
      containers:
      - image: nginx:1.16
        name: nginx
```

由此可见，只需要给kustomize build命令传递不同的kustomization目录路径，就可以得到相对应的配置渲染输出。

9.7 本章小结

本章主要介绍了Kustomize的相关概念、用法等，但是无法详尽地介绍Kustomize的所有特性。希望通过本章的学习，能够帮助大家在Operator项目中更容易看懂Kustomize相关配置。当然，如果大家对Kustomize感兴趣，想进一步深入学习Kustomize，可以仔细阅读一下其官网文档。

第 10 章 使用 Helm 打包应用

Helm是一个Kubernetes包管理工具，就像Linux系列的Yum和Apt一样，Helm通过chart的方式组织Kubernetes之上的应用资源。我们完成一个云原生应用的业务逻辑编码后，需要部署到Kubernetes集群中，繁多的YAML配置编写和维护过程是比较枯燥的，使用Helm能够很大程度上降低Kubernetes之上微服务部署的复杂度，提高工作效率。

Helm几乎是发布云原生应用绕不开的工具，当然也包括我们编写的Operator应用，所以有必要好好学习Helm这个工具。

10.1 Helm 的安装

在Helm的release页面https://github.com/helm/helm/releases可以看到当前最新的发布版本，如图10-1所示。

Installation and Upgrading

Download Helm v3.8.0. The common platform binaries are here:

- MacOS amd64 (checksum / 532ddd6213891084873e5c2dcafa577f425ca662a6594a3389e288fc48dc2089)
- MacOS arm64 (checksum / 751348f1a4a876ffe089fd68df6aea310fd05fe3b163ab76aa62632e327122f3)
- Linux amd64 (checksum / 8408c91e846c5b9ba15eb6b1a5a79fc22dd4d33ac6ea63388e5698d1b2320c8b)
- Linux arm (checksum / 05e900d0688edd8d455e0d4c51b419cd2b10120d485be7a1262582f51c92e941)
- Linux arm64 (checksum / 23e08035dc0106fe4e0bd85800fd795b2b9ecd9f32187aa16c49b0a917105161)
- Linux i386 (checksum / ed845e2e2631eddf526cab606f023fded03c5b99175f7cacd5edcd41800c7e45)
- Linux ppc64le (checksum / 5070fa5188e7bc798dd54bc1ea9fc4cda623d9ff45eedb05ec93db234309f391)
- Linux s390x (checksum / f8088ea57290fcc0aae50e2075c1adc258247cdf55169e9a9ca7762a64f558db)
- Windows amd64 (checksum / d52e0cda6c4cc0e0717d5161ca1ba7a8d446437afdbe42b3c565c145ac752888)

图 10-1 查看 Helm 的版本

用户可以选择下载自己系统对应的Helm版本，在自己的环境中下载解压后会得到一个darwin-arm64目录，接着将darwin-arm64中的Helm文件存放到合适的位置：

```
# sudo mv darwin-arm64/helm /usr/local/bin
# helm version
version.BuildInfo{Version:"v3.8.0", GitCommit:
"d14138609b01886f544b2025f5000351c9eb092e",GitTreeState:"clean", GoVersion:"go1.17.5"}
```

10.2　Helm 的基本概念

1. Chart

一个Chart指的是一个Helm包，类似Yum的rpm包或者Apt的dpkg包，涵盖一个软件安装所需的一切"物料"，Chart中包含的是在Kubernetes中运行的一个云原生应用方方面面的资源配置。

2. Repository

知道了Chart包的概念之后，肯定很快就能想到这些Chart应该存放到哪里。没错，用来放置和分发Chart包的地方就叫Repository，即Chart仓库的意思。

3. Release

一个Chart包可以在同一个Kubernetes集群中被多次部署，每次部署产生的实例就叫作一个Release。比如同样一个Web应用打包成Chart之后，可以在dev和test两个namespace中分别部署一份实例，出于不同的目的，这里的两个实例就是两个Release。

4. ArtifactHub

类似于DockerHub和GitHub，Helm的Chart也有一个Hub，其地址是https://artifacthub.io。当然，ArtifactHub中存放的不只是Chart，还有kubectl plugins、OLM operators、Tekton tasks等，不过目前我们只关心Chart部分，如图10-2所示。

图 10-2　ArtifactHub

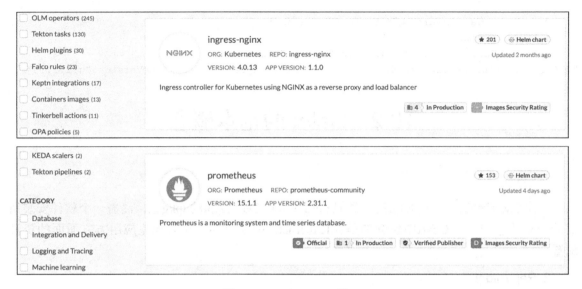

图 10-2　ArtifactHub（续）

可以看到左上角有一个KIND可选类型，我们选择Helm charts就能看到所有托管在这里的Chart了。

后面讲解Helm的命令的时候会再详细介绍怎么使用ArtifactHub。

10.3　Helm 的常用操作

本节学习Helm的一些常用的基础命令。

10.3.1　搜索 Chart 包

本小节主要用到下面几个命令：

- helm search hub [KEYWORD] [flags]：在ArtifactHub中从所有的Repository内搜索相关Chart。
- helm repo add [NAME] [URL] [flags]：本地添加一个Repository。
- helm repo list [flags]：列出本地所有Repository，list可以简写为ls。
- helm search repo [keyword] [flags]：在本地Repository搜索匹配的Chart。

安装好Helm客户端命令后，可以通过search命令来找到自己想要的Chart包。以kube-prometheus-stack为例，可以在ArtifactHub中搜索相应的Chart：

```
helm search hub kube-prometheus-stack
```

由于输出结果太宽，可能大家会看到如图10-3所示的结果。

从这里很难区分前两行输出的URL，可以通过--max-col-width来指定输出宽度，比如把命令改成如下：

```
helm search hub kube-prometheus-stack --max-col-width=100
```

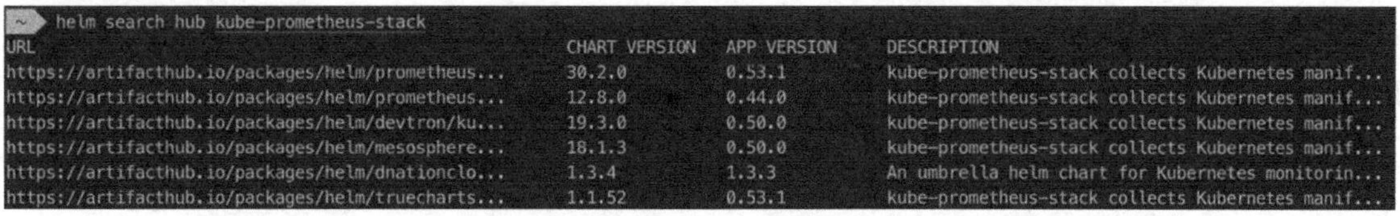

图 10-3 查看结果

这时URL就能显示完整，如图10-4所示。

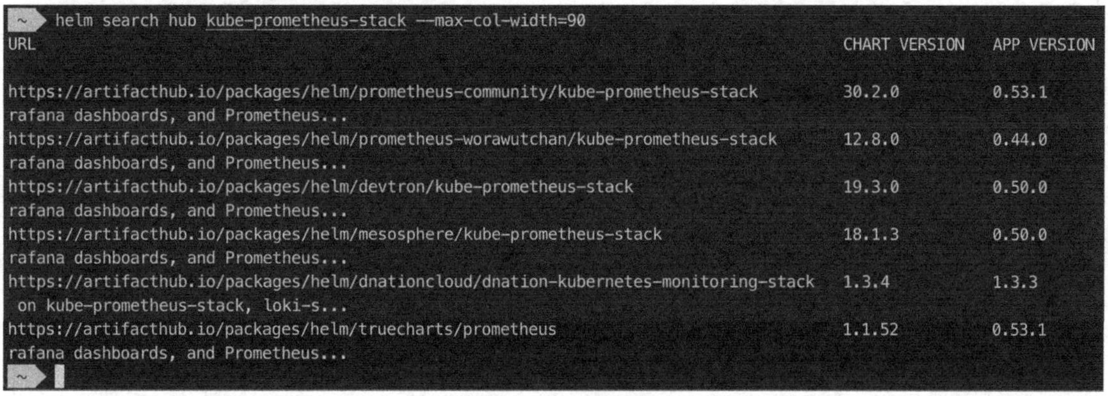

图 10-4 显示完整的 URL

可以看到第一个URL才是Prometheus社区官方提供的链接，然后需要打开浏览器输入这个URL：https://artifacthub.io/packages/helm/prometheus-community/kube-prometheus-stack，接着可以看到如图10-5所示的页面。

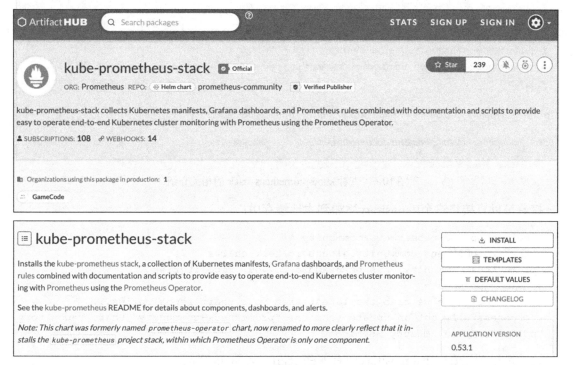

图 10-5 kube-prometheus-stack 页面

这里是kube-prometheus-stack的相关介绍，继续往下翻页可以找到相应Repository的地址、前置条件、安装方式和依赖的Chart等信息，如图10-6所示。

Prerequisites

- Kubernetes 1.16+
- Helm 3+

Get Repo Info

```
helm repo add prometheus-community https://prometheus-community.github.io/helm-charts
helm repo update
```

See helm repo for command documentation.

Install Chart

```
# Helm
$ helm install [RELEASE_NAME] prometheus-community/kube-prometheus-stack
```

See configuration below.

See helm install for command documentation.

Dependencies

By default this chart installs additional, dependent charts:

- prometheus-community/kube-state-metrics
- prometheus-community/prometheus-node-exporter
- grafana/grafana

To disable dependencies during installation, see multiple releases below.

See helm dependency for command documentation.

图 10-6　查看 kube-prometheus-stack 的相关介绍

接着可以直接将这个Repository添加到本地缓存中：

```
# helm repo add prometheus-community \
https://prometheus-community.github.io/helm-charts
"prometheus-community" has been added to your repositories
# helm repo update
Hang tight while we grab the latest from your chart repositories...
...Successfully got an update from the "prometheus-community" chart repository
Update Complete. ⎈Happy Helming!⎈
```

这时可以顺便看一下Repository中有哪些Chart，如图10-7所示。

```
~ helm repo list
NAME                    URL
prometheus-community    https://prometheus-community.github.io/helm-charts

~ helm search repo prometheus-community
NAME                                              CHART VERSION  APP VERSION  DESCRIPTION
prometheus-community/alertmanager                 0.14.0         v0.23.0      The Alertmanager handles alert
prometheus-community/kube-prometheus-stack        30.2.0         0.53.1       kube-prometheus-stack collects
prometheus-community/kube-state-metrics           4.4.1          2.3.0        Install kube-state-metrics to
prometheus-community/prometheus                   15.1.1         2.31.1       Prometheus is a monitoring sys
prometheus-community/prometheus-adapter           3.0.1          v0.9.1       A Helm chart for k8s prometheu
prometheus-community/prometheus-blackbox-exporter 5.3.1          0.19.0       Prometheus Blackbox Exporter
```

图 10-7　查看存放的 Chart

目前这个Repository中存放了27个Chart，基本包含Prometheus相关的各种工具。

10.3.2　安装 Chart 包

1. 通过Helm安装一个kube-prometheus-stack实例

有了Chart包之后，下一步是安装这个包，所以本小节继续学习各种安装Chart包的方式。安装命令的格式如下：

```
helm install [NAME] [CHART] [flags]
```

NAME指的是Instance的名字。CHART也就是要使用的Chart包地址，既可以是远程仓库中的包，也可以是本地的Chart压缩包，或者是解压后的Chart包，我们具体来看这几种用法。

以kube-prometheus-stack的安装为例：

```
# helm install prometheus-community/kube-prometheus-stack \
-n monitoring -g
NAME: kube-prometheus-stack-1643463695
LAST DEPLOYED: Sat Jan 29 21:41:40 2022
NAMESPACE: monitoring
STATUS: deployed
REVISION: 1
NOTES:
kube-prometheus-stack has been installed. Check its status by running:
  kubectl --namespace monitoring get pods -l "release=kube-prometheus-stack-
1643463695"

Visit https://github.com/prometheus-operator/kube-prometheus for instructions on
how to create & configure Alertmanager and Prometheus instances using the Operator.
```

这里用的命令是helm install prometheus-community/kube-prometheus-stack -n monitoring -g，先看每个参数的含义：

- prometheus-community/kube-prometheus-stack：表示Chart地址，也就是在本地仓库prometheus-community中的kube-prometheus-stack。
- -n monitoring：等价于--namespace monitoring，表示这个实例将安装在monitoring 命名空间下，如果这个命名空间不存在，则会被自动创建。
- -g：等价于--generate-name，表示不手动指定实例名，而是自动生成一个名字来使用。

接着查看monitoring命名空间中的Pod是否都运行起来了：

```
# kubectl get pod -n monitoring
NAME                                                              READY   STATUS    RESTARTS   AGE
alertmanager-kube-prometheus-stack-1643-alertmanager-0            2/2     Running   0          11s
kube-prometheus-stack-1643-operator-6c9fbb58cc-xk4cf              1/1     Running   0          15s
kube-prometheus-stack-1643463695-grafana-56b8d79f9-mzl7p          3/3     Running   0          15s
kube-prometheus-stack-1643463695-kube-state-metrics-5df9d527fss   1/1     Running   0          15s
kube-prometheus-stack-1643463695-prometheus-node-exporter-hd5fx   1/1     Running   0          13s
prometheus-kube-prometheus-stack-1643-prometheus-0                1/2     Running   0          10s
```

可以看到prometheus、prometheus-operator、alertmanager、grafana、kube-state-metrics和node-exporter等Pod都已经运行起来了。这里除了Pod外，还有相应的Deployment、DaemonSet、StatefulSet、Service以及各种CRDs被创建出来。想象一下如果不使用Helm Chart的方式来组织这些资源文件，将要花费多少力气去寻找和梳理这么多的资源配置。

Helm默认不会等待所有Pod都运行起来，因为很多应用的镜像拉取会比较耗时，甚至超时失败。在使用helm install命令退出后，还可以通过helm ls来列出所有已经安装的Chart，也可以通过helm status来查看某个Chart实例的状态。

使用helm ls -n monitoring命令找到刚才部署的实例名：

```
# helm ls -n monitoring
NAME                               NAMESPACE   REVISION  UPDATED                              STATUS    CHART                         APP VERSION
kube-prometheus-stack-1643463695   monitoring  1         2022-01-29 21:41:40.387995 +0800 CST deployed  kube-prometheus-stack-30.2.0  0.53.1
```

使用helm status kube-prometheus-stack-1643463695 -n monitoring命令查看某个实例的状态：

```
# helm status kube-prometheus-stack-1643463695 -n monitoring
NAME: kube-prometheus-stack-1643463695
LAST DEPLOYED: Sat Jan 29 21:41:40 2022
NAMESPACE: monitoring
STATUS: deployed
REVISION: 1
NOTES:
kube-prometheus-stack has been installed. Check its status by running:
  kubectl --namespace monitoring get pods -l "release=kube-prometheus-stack-1643463695"

Visit https://github.com/prometheus-operator/kube-prometheus for instructions on how to create & configure Alertmanager and Prometheus instances using the Operator.
```

2. Helm安装资源的顺序

Helm安装各种Kubernetes是遵循一定顺序的，比如Namespace和Deployment都需要创建时，先创建后者肯定就走不通了。这个顺序是：

1）Namespace
2）NetworkPolicy
3）ResourceQuota
4）LimitRange
5）PodSecurityPolicy
6）PodDisruptionBudget
7）ServiceAccount
8）Secret
9）SecretList
10）ConfigMap
11）StorageClass
12）PersistentVolume
13）PersistentVolumeClaim
14）CustomResourceDefinition
15）ClusterRole
16）ClusterRoleList
17）ClusterRoleBinding
18）ClusterRoleBindingList
19）Role
20）RoleList
21）RoleBinding
22）RoleBindingList
23）Service
24）DaemonSet
25）Pod
26）ReplicationController
27）ReplicaSet
28）Deployment
29）HorizontalPodAutoscaler
30）StatefulSet
31）Job
32）CronJob
33）Ingress
34）APIService

3. 其他安装方法

除了直接使用本地Repository中的Chart索引来安装一个实例外，我们还有其他3种方法来完成安装过程。这4种方法分别是：

- helm install chartrepo/chartname：直接从Repository安装。
- helm install ./chartname-1.2.3.tgz：通过helm pull下载。
- helm install ./chartname：解压这个压缩包。
- helm install https://chartrepo.com/charts/chartname-1.2.3.tgz：从一个远程地址安装。

10.3.3 自定义 Chart 配置

前面通过helm install命令部署了一个kube-prometheus-stack实例，大家可能已经注意到没有对这个实例进行任何配置。大多数情况下，需要知道自己部署的应用支持哪些配置项，然后根据具体的应用场景去调整相应的配置，比如部署到开发环境时分配更小的内存和实例数，部署到生产环境需要更大的内存和实例数分配。

1. helm show values

怎么查看一个Chart的配置呢？首先可以通过helm show values命令：

```
# helm show values prometheus-community/kube-prometheus-stack
...
nameOverride: ""
namespaceOverride: ""
kubeTargetVersionOverride: ""
```

```
kubeVersionOverride: ""
...
```

大家看到的输出会和这里写得不一样,因为笔者删除了注释部分,并且只保留了几行配置项,实际这个values配置多达近3000行。所以helm show values更加适用于配置比较简单的Chart。

2. helm pull

为了看清一个配置复杂的Chart包的全貌,需要将这个Chart下载下来,然后解压缩包中的配置文件:

```
# helm pull prometheus-community/kube-prometheus-stack
# ls
kube-prometheus-stack-30.2.0.tgz
# tar -xzvf kube-prometheus-stack-30.2.0.tgz
x kube-prometheus-stack/Chart.yaml
x kube-prometheus-stack/Chart.lock
x kube-prometheus-stack/values.yaml
x kube-prometheus-stack/templates/NOTES.txt
x kube-prometheus-stack/templates/_helpers.tpl
x kube-prometheus-stack/templates/alertmanager/alertmanager.yaml
...
```

这个Chart包下载下来后的名字是kube-prometheus-stack-30.2.0.tgz,解压后得到一个kube-prometheus-stack目录,这个目录内有如下的文件/目录:

```
# ll kube-prometheus-stack
total 280
-rw-r--r--   1 danielhu  wheel   656B  1 25 21:34 CONTRIBUTING.md
-rw-r--r--   1 danielhu  wheel   450B  1 25 21:34 Chart.lock
-rw-r--r--   1 danielhu  wheel   1.7K  1 25 21:34 Chart.yaml
-rw-r--r--   1 danielhu  wheel    35K  1 25 21:34 README.md
drwxr-xr-x   5 danielhu  wheel   160B  1 29 22:58 charts
drwxr-xr-x  10 danielhu  wheel   320B  1 29 22:58 crds
drwxr-xr-x   9 danielhu  wheel   288B  1 29 22:58 templates
-rw-r--r--   1 danielhu  wheel    90K  1 25 21:34 values.yaml
```

打开values.yaml,可以看到其中有一个prometheus.prometheusSpec.replicas配置项,我们以这个配置为例,学习如何在部署时自定义配置。

3. 自定义配置内容

具体操作之前,为了排除干扰,下面每次部署执行前大家记得清理已经部署好的Chart实例,可以用helm ls -n monitoring命令查看实例名,然后通过helm uninstall -n monitoring [Release Name]来卸载。

方法一:直接将需要自定义的配置写入新文件。

```
# cat <<EOF > values-replicas.yaml
prometheus:
  prometheusSpec:
    replicas: 2
```

```
EOF
# helm install prometheus-community/kube-prometheus-stack -n monitoring -g -f
values-replicas.yaml
NAME: kube-prometheus-stack-1643469058
LAST DEPLOYED: Sat Jan 29 23:11:08 2022
NAMESPACE: monitoring
STATUS: deployed
REVISION: 1
NOTES:
kube-prometheus-stack has been installed. Check its status by running:
   kubectl --namespace monitoring get pods -l
"release=kube-prometheus-stack-1643469058"

Visit https://github.com/prometheus-operator/kube-prometheus for instructions on
how to create & configure Alertmanager and Prometheus instances using the Operator.
```

在安装时通过-f参数指定一个新的values-replicas.yaml文件来自定义配置内容，查看一下效果：

```
# kubectl get pod -n monitoring \
-l app.kubernetes.io/name=prometheus
NAME                                                  READY   STATUS    RESTARTS   AGE
prometheus-kube-prometheus-stack-1643-prometheus-0    2/2     Running   0          2m11s
prometheus-kube-prometheus-stack-1643-prometheus-1    2/2     Running   0          2m11s
```

Prometheus的副本数确实变成2了。

方法二：通过--set来指定自定义配置项。

```
# helm install prometheus-community/kube-prometheus-stack -n monitoring -g --set
prometheus.prometheusSpec.replicas=2
NAME: kube-prometheus-stack-1643470203
LAST DEPLOYED: Sat Jan 29 23:30:05 2022
NAMESPACE: monitoring
STATUS: deployed
REVISION: 1
NOTES:
kube-prometheus-stack has been installed. Check its status by running:
   kubectl --namespace monitoring get pods -l
"release=kube-prometheus-stack-1643470203"

Visit https://github.com/prometheus-operator/kube-prometheus for instructions on
how to create & configure Alertmanager and Prometheus instances using the Operator.
```

这时Prometheus的副本数也会被成功改为2。

```
# kubectl get pod -n monitoring \
-l app.kubernetes.io/name=prometheus
NAME                                                  READY   STATUS    RESTARTS   AGE
prometheus-kube-prometheus-stack-1643-prometheus-0    2/2     Running   0          119s
prometheus-kube-prometheus-stack-1643-prometheus-1    2/2     Running   0          119s
```

在需要自定义的配置项不太多的时候，--set是简洁高效的，反之一定需要准备一个values.yaml文件。

--set格式和YAML配置的对应关系大致如表10-1所示。

表 10-1 --set 格式和 YAML 配置的对应关系

--set 格式	YAML 配置
--set key=value	key: value
--set key=value, key1=value1	key: value key1: value1
--set outer.inner=xxx	outer: inner: xxx
--set key={a, b, c}	key: - a - b - c
--set key[0].a=123	key: - a: 123
--set key=value1\,value2	key: "value1,value2"
--set a\.b=xxx	a.b: xxx

10.3.4 Release 升级与回滚

当一个Chart发布了新版本，或者想要更新同一个版本Chart包的一个实例时，可以通过helm upgrade命令来完成。如果更新之后需要回滚，则可以对应使用helm rollback命令。不知道想要回滚到哪个版本，就使用helm history命令。本小节涉及的一些新命令具体格式如下：

- helm upgrade [RELEASE] [CHART] [flags]：更新一个Chart实例。
- helm history RELEASE_NAME [flags]：打印一个Release的所有历史修订版本（Revisions）。
- helm rollback <RELEASE> [REVISION] [flags]：回滚一个Release到指定版本。

1. 升级

以10.3.3节自定义Chart配置时使用的2个副本Prometheus的kube-prometheus-stack为例，通过helm upgrade来将其更新为3个副本。

先部署好2个副本Prometheus的kube-prometheus-stack，具体配置参见10.3.3节：

```
# helm install ./kube-prometheus-stack -n monitoring -g -f values-replicas.yaml
NAME: kube-prometheus-stack-1643553862
LAST DEPLOYED: Sun Jan 30 22:44:23 2022
NAMESPACE: monitoring
STATUS: deployed
REVISION: 1
NOTES:
kube-prometheus-stack has been installed. Check its status by running:
  kubectl --namespace monitoring get pods -l "release=kube-prometheus-stack-1643553862"
```

Visit https://github.com/prometheus-operator/kube-prometheus for instructions on how to create & configure Alertmanager and Prometheus instances using the Operator.

这时的Pod状态如下：

```
# kubectl get pod -n monitoring
NAME                                                              READY   STATUS    RESTARTS   AGE
alertmanager-kube-prometheus-stack-1643-alertmanager-0            2/2     Running   0          98s
kube-prometheus-stack-1643-operator-5d96d8f9df-vn8kf              1/1     Running   0          100s
kube-prometheus-stack-1643553862-grafana-57cbc68d77-gj6f8         3/3     Running   0          100s
kube-prometheus-stack-1643553862-kube-state-metrics-7564fbphgf6   1/1     Running   0          100s
kube-prometheus-stack-1643553862-prometheus-node-exporter-kpl65   1/1     Running   0          100s
prometheus-kube-prometheus-stack-1643-prometheus-0                2/2     Running   0          98s
prometheus-kube-prometheus-stack-1643-prometheus-1                2/2     Running   0          98s
```

然后修改配置如下：

```
# cat <<EOF > values-replicas.yaml
prometheus:
  prometheusSpec:
    replicas: 3
EOF
```

开始升级Release：

```
# helm upgrade kube-prometheus-stack-1643553862 ./kube-prometheus-stack -n monitoring -f values-replicas.yaml
Release "kube-prometheus-stack-1643553862" has been upgraded. Happy Helming!
NAME: kube-prometheus-stack-1643553862
LAST DEPLOYED: Sun Jan 30 22:49:43 2022
NAMESPACE: monitoring
STATUS: deployed
REVISION: 2
NOTES:
kube-prometheus-stack has been installed. Check its status by running:
  kubectl --namespace monitoring get pods -l "release=kube-prometheus-stack-1643553862"
```

Visit https://github.com/prometheus-operator/kube-prometheus for instructions on how to create & configure Alertmanager and Prometheus instances using the Operator.

可以看到这时REVISION更新成2。我们继续查看Release列表：

```
# helm ls -n monitoring
NAME                    NAMESPACE   REVISION   UPDATED    STATUS   CHART           APP VERSION
```

```
kube-prometheus-stack-1643553862  monitoring    2           2022-01-30 22:49:43.618514
+0800 CST    deployed kube-prometheus-stack-30.2.0  0.53.1
```

也可以通过helm history命令查看历史Release信息：

```
# helm history kube-prometheus-stack-1643553862 -n monitoring
REVISION UPDATED                  STATUS     CHART                          APP
VERSION  DESCRIPTION
1        Sun Jan 30 22:44:23 2022 superseded kube-prometheus-stack-30.2.0
0.53.1            Install complete
2        Sun Jan 30 22:49:43 2022 deployed   kube-prometheus-stack-30.2.0
0.53.1            Upgrade complete
```

最后看Pod是否真的更新了：

```
kubectl get pod -n monitoring
NAME                                                            READY   STATUS
RESTARTS   AGE
alertmanager-kube-prometheus-stack-1643-alertmanager-0           2/2     Running
0          8m28s
kube-prometheus-stack-1643-operator-5d96d8f9df-vn8kf             1/1     Running
0          8m30s
kube-prometheus-stack-1643553862-grafana-57cbc68d77-gj6f8        3/3     Running
0          8m30s
kube-prometheus-stack-1643553862-kube-state-metrics-7564fbphgf6  1/1
Running    0          8m30s
kube-prometheus-stack-1643553862-prometheus-node-exporter-kpl65  1/1
Running    0          8m30s
prometheus-kube-prometheus-stack-1643-prometheus-0               2/2     Running
0          8m28s
prometheus-kube-prometheus-stack-1643-prometheus-1               2/2     Running
0          8m28s
prometheus-kube-prometheus-stack-1643-prometheus-2               2/2     Running
0          3m10s
```

从这里大家或许注意到一个细节，就是新增的Pod单独被创建出来了，其他资源没有发生变化。也就是说，helm upgrade的过程其实是最小化变更发生了变化的资源，而不是推翻重建所有资源对象实例。接着看如何回滚到上一个版本。

2. 回滚

刚才通过helm history命令看到当前kube-prometheus-stack-1643553862实例有两个 Revision，我们尝试将其回滚到Revision 1：

```
# helm rollback kube-prometheus-stack-1643553862 1 -n monitoring
Rollback was a success! Happy Helming!
```

这里没有太多的日志输出，我们查看现在helm history命令会带来哪些信息：

```
# helm history kube-prometheus-stack-1643553862 -n monitoring
REVISION UPDATED                  STATUS     CHART                          APP
VERSION  DESCRIPTION
1        Sun Jan 30 22:44:23 2022 superseded kube-prometheus-stack-30.2.0
```

```
                  0.53.1           Install complete
2                 Sun Jan 30 22:49:43 2022 superseded   kube-prometheus-stack-30.2.0
                  0.53.1           Upgrade complete
3                 Sun Jan 30 23:04:15 2022 deployed    kube-prometheus-stack-30.2.0
                  0.53.1           Rollback to 1
```

笔者把新增的一行加粗了，可以看到这时多了第3个Revision，描述信息是"Rollback to 1"，也就是说Helm并没有直接抛弃2版本，直接回滚到1版本，而是新增了一个3版本，只是配置和1版本完全一致。这也非常合理，不然怎么回滚到2版本呢？

这时查看Pod列表，也可以很和谐地看到新增的那个Pod被删除了：

```
# kubectl get pod -n monitoring
   NAME                                                                 READY   STATUS
RESTARTS    AGE
   alertmanager-kube-prometheus-stack-1643-alertmanager-0               2/2     Running
0           23m
   kube-prometheus-stack-1643-operator-5d96d8f9df-vn8kf                 1/1     Running
0           23m
   kube-prometheus-stack-1643553862-grafana-57cbc68d77-gj6f8            3/3     Running
0           23m
   kube-prometheus-stack-1643553862-kube-state-metrics-7564fbphgf6      1/1
Running    0        23m
   kube-prometheus-stack-1643553862-prometheus-node-exporter-kpl65      1/1
Running    0        23m
   prometheus-kube-prometheus-stack-1643-prometheus-0                   2/2     Running
0           23m
   prometheus-kube-prometheus-stack-1643-prometheus-1                   2/2     Running
0           23m
```

10.3.5　Release 卸载

前面其实已经用过helm uninstall命令了，不过没有太详细介绍，本小节再仔细学习helm uninstall的用法。

```
helm uninstall RELEASE_NAME [...] [flags]
```

毫无疑问，如果执行helm uninstall kube-prometheus-stack-1643553862，则会把kube-prometheus-stack-1643553862这个实例卸载。下面介绍另一种卸载方式：

```
# helm uninstall kube-prometheus-stack-1643553862 -n monitoring --keep-history
release "kube-prometheus-stack-1643553862" uninstalled
```

当然，kube-prometheus-stack-1643553862这个实例也被成功卸载了，但是我们仍然可以通过helm history命令看到历史版本记录：

```
# helm history kube-prometheus-stack-1643553862 -n monitoring
   REVISION UPDATED                    STATUS       CHART                         APP
VERSION  DESCRIPTION
   1        Sun Jan 30 22:44:23 2022   superseded   kube-prometheus-stack-30.2.0
0.53.1             Install complete
```

```
2              Sun Jan 30 22:49:43 2022  superseded   kube-prometheus-stack-30.2.0
0.53.1         Upgrade complete
3              Sun Jan 30 23:04:15 2022  uninstalled  kube-prometheus-stack-30.2.0
0.53.1         Uninstallation complete
```

看到这里也许已经想到，如果我们执行helm rollback命令，是否能够让这个实例"起死回生"呢？试一下吧：

```
# helm rollback kube-prometheus-stack-1643553862 2 -n monitoring
Rollback was a success! Happy Helming!
# helm history kube-prometheus-stack-1643553862 -n monitoring
REVISION  UPDATED                   STATUS        CHART                          APP
VERSION   DESCRIPTION
1              Sun Jan 30 22:44:23 2022  superseded   kube-prometheus-stack-30.2.0
0.53.1         Install complete
2              Sun Jan 30 22:49:43 2022  superseded   kube-prometheus-stack-30.2.0
0.53.1         Upgrade complete
3              Sun Jan 30 23:04:15 2022  uninstalled  kube-prometheus-stack-30.2.0
0.53.1         Uninstallation complete
4              Sun Jan 30 23:16:28 2022  deployed     kube-prometheus-stack-30.2.0
0.53.1         Rollback to 2
```

helm rollback执行成功了，history记录中也多了4版本，看起来没有任何问题，这个实例成功恢复了。

这时查看一下Pod列表，可以看到新创建的Pod都运行起来了：

```
# kubectl get pod -n monitoring
NAME                                                              READY   STATUS
RESTARTS   AGE
  alertmanager-kube-prometheus-stack-1643-alertmanager-0          2/2     Running
0          3m33s
  kube-prometheus-stack-1643-operator-5d96d8f9df-7rhmt            1/1     Running
0          3m35s
  kube-prometheus-stack-1643553862-grafana-57cbc68d77-8j2bj       3/3     Running
0          3m35s
  kube-prometheus-stack-1643553862-kube-state-metrics-7564fbxqccm  1/1
Running   0          3m35s
  kube-prometheus-stack-1643553862-prometheus-node-exporter-27h8t  1/1
Running   0          3m35s
  prometheus-kube-prometheus-stack-1643-prometheus-0              2/2     Running
0          3m33s
  prometheus-kube-prometheus-stack-1643-prometheus-1              2/2     Running
0          3m33s
  prometheus-kube-prometheus-stack-1643-prometheus-2              2/2     Running
0          3m33s
```

10.3.6 Helm 命令的常用参数

Helm的install/upgrade/rollback等子命令都有几个很有用的参数，本小节将介绍其中的两个参数。

- --timeout：等待Kubernetes命令执行完成的超时时间，默认是50毫秒，这里的值是一个Golang的Duration。
- --wait：等待所有的Pod变成准备（ready）状态，PVC完成绑定，deployments至少有（Desired Pods 减去maxUnavailable Pods）数量处于准备状态，Service都有IP成功绑定。如果到了--timeout指定的超时时间，那么这个Release就会被标记为FAILED状态。

10.4 封装自己的 Chart 包

Helm将描述一个应用如何将Kubernetes之上部署的各种相关资源文件打包在一起，这个打包格式或者这个包叫作Chart。Chart就是一个包含很多文件的目录，这个目录可以被打成一个有版本的压缩包，也可以被部署。本节详细学习Chart的格式，以及如何封装自己的Chart包。

10.4.1 Chart 的目录结构

还是以kube-prometheus-stack为例，通过helm pull命令下载这个Chart包，然后解压缩，看其中的目录结构。

```
# helm pull prometheus-community/kube-prometheus-stack
# tar -xzvf kube-prometheus-stack-30.2.0.tgz
x kube-prometheus-stack/Chart.yaml
x kube-prometheus-stack/Chart.lock
x kube-prometheus-stack/values.yaml
x kube-prometheus-stack/templates/NOTES.txt
x kube-prometheus-stack/templates/_helpers.tpl
...
```

解压缩后会得到一个kube-prometheus-stack目录，也就是这个Chart的根目录，与Chart的名字相同，但是没有版本号信息。

kube-prometheus-stack中主要有如下的文件/目录：

- Chart.yaml：包含Chart基本信息的YAML文件。
- README.md：可选的readme文件。
- CONTRIBUTING.md：可选的贡献者指引文件。
- Chart.lock：当前Chart的依赖锁定文件。
- values.yaml：当前Chart的默认配置文件。
- charts/：这个目录中放置的是当前Chart依赖的所有Chart。
- crds/：自定义资源配置存放目录，也就是当前Chart相关的所有CRD资源。
- templates/：存放模板文件的目录，模板与values配置加在一起可以渲染成完整的Kubernetes配置文件。
- templates/NOTES.txt：可选的纯文本使用帮助信息，也就是完成安装操作后控制台输出的帮助信息文本。

10.4.2 Chart.yaml 文件

Chart.yaml是一个必选文件，其中存放的是一个Chart的基本描述信息，格式如下：

```
apiVersion:             // Chart的API版本（必选）
name:                   // Chart的名字（必选）
version:                // 一个SemVer 2格式的版本信息（必选）
kubeVersion:            // 一个用来描述兼容的Kubernetes版本的SemVer格式范围信息（可选）
description:            // 当前项目的一句话描述（可选）
type:                   // 当前Chart的类型（可选）
keywords:
  -                     // 当前项目的一些关键字（可选）
home:                   // 当前项目的主页URL（可选）
sources:
  -                     // 当前项目的源码URL列表（可选）
dependencies:           // 当前项目的依赖列表（可选）
  - name:               // 依赖的Chart的名字
    version:            // 依赖的Chart的版本
    repository:         // 依赖的Chart代码库的URL，比如https://example.com/charts或者
"@repo-name" 这种简写格式
    condition:          // 一个可选的YAML配置路径，需要是bool值，用来表示指定当前Chart是否启
用，比如kube-prometheus-stack中的grafana.enabled p配置（可选）
    tags:               // 可选的tag配置
      -                 // tag的作用是可以将这些依赖Chart分组，用于实现批量启用/禁用
    import-values:      // 可选
    alias:              // 当前Chart的别名，主要用在一个Chart需要被多次添加的时候
maintainers:            // 可选
  - name:               // 维护者姓名
    email:              // 维护者电子邮箱
    url:                // 维护者个人站点URL
icon:                   // 一个SVG或者PNG 格式的图标图片URL（可选）
appVersion:             // 可选的应用版本信息，不必是SemVer格式（可选）
deprecated:             // 标记当前Chart是否废弃，用bool值（可选）
annotations:            // 目前Helm已经不允许在Chart.yaml中添加额外的配置项，如果需要额外的
自定义配置，只能添加在注解中
    example:            // 注解列表（可选）
```

每个Chart需要有明确的版本信息，Version在kube-prometheus-stack中的当前值是30.2.0，对应压缩包名字是kube-prometheus-stack-30.2.0.tgz，也就是Chart.yaml中的Version字段会被用在Chart包的名字中。这个Version的格式是SemVer 2，SemVer的具体定义可以在https://semver.org/spec/v2.0.0.html中查阅。SemVer也就是Semantic Versioning的简写，翻译过来就是"语义版本"的意思。我们平时经常看到的x.y.z格式的版本号，也就是MAJOR.MINOR.PATCH（主版本号.小版本号.补丁版本号）格式，就是SemVer版本格式。

apiVersion字段目前都是配置成v2版本，在以前也用过v1版本，虽然Helm3目前也能识别v1版本，但是除非有不好绕过的历史兼容性负担，否则我们都使用v2版本。

另外，appVersion和Version看着容易混淆。appVersion描述的是应用本身的版本，比如我们用的kube-prometheus-stack-30.2.0.tgz的appVersion是0.53.1。appVersion更多的是进行"信息记录"，

不会影响这个Chart本身的版本计算，也就是说假如你写错了，只是容易误导人，并不太会影响Chart成功部署。

type表示这个Chart的类型，这里的类型默认是application，另一个可选值是library。

kubeVersion字段用来描述兼容的Kubernetes版本信息，这个字段会被Helm识别，并且在部署时进行兼容性校验。这里有一些规则需要了解：

- >= 1.10.0 < 1.23.0：这种记录会被解析成Kubernetes版本，需要不小于1.10.0且小于1.23.0。
- >= 1.10.0 < 1.20.0 || >= 1.20.1 <= 1.23.0："或"的含义，可以通过"||"操作符来描述，这种写法表示Kubernetes版本在除了1.20.0之外的[1.10.0, 1.23.0]之间。
- >= 1.10.0 <= 1.23.0 != 1.20.0：上面这种排除1.20.0版本的方式可以用更简单的方式来描述，就是这里的!=。
- =、!=、<、>、<=、>=：这些操作符都可以用。
- ~1.2.3：表示补丁版本可以随意选择，也就是>= 1.2.3 <= 1.3.0的意思。
- ^1.2.3：表示小版本号可以随意选择，也就是>= 1.2.3 <= 2.0.0的意思。

10.4.3　Chart 依赖管理

一个Chart和一个普通的Go语言项目一样，绕不开依赖管理的问题。一个Chart可以依赖N个其他Chart，这些依赖Chart可以动态链接到当前Chart中，通过定义在Chart.yaml中的dependencies字段。另外，也可以直接将依赖的Chart直接放置到chart/目录下，静态管理这些依赖Chart。

前面介绍Chart.yaml文件时已经提到过dependencies字段了，大致含义是这样的：

```
dependencies:            // 当前项目的依赖列表（可选）
  - name:                // 依赖的Chart的名字
    version:             // 依赖的Chart的版本
    repository:          // 依赖的Chart代码库的URL，比如https://example.com/charts或者
"@repo-name"这种简写格式
    condition:           // 一个可选的YAML配置路径，需要是bool值，用来表示指定当前Chart是否启
用，比如kube-prometheus-stack中的grafana.enabled p配置（可选）
    tags:                // 可选的tag配置
      -                  // tag 的作用是可以将这些依赖Chart分组，用于实现批量启用/禁用
    import-values:       // 可选
    alias:               // 当前Chart的别名，主要用于一个Chart需要被多次添加的时候
```

这里的repository是一个依赖Chart的代码库URL，这个URL也是需要通过helm repo add命令添加到本地Repository列表中的，这时同样可以通过helm repo add命令使用的NAME来替换URL填入dependencies[x].repository中。

当Chart.yaml中定义好dependencies之后，就可以通过helm dependency update命令将所有的依赖Chart下载到本地的chart/目录下。

- helm dependency update CHART [flags]：根据Chart.yaml文件中的依赖定义更新本地依赖。
- helm dependency list CHART [flags]：列出当前Chart的所有依赖。
- helm dependency build CHART [flags]：从Chart.lock重建本地charts/下的本地依赖。

另外，dependency子命令有点长，其实可以简写为dep，update可以简写为up，list可以简写为ls。

10.5 本章小结

本章主要介绍了Helm的基本概念、常用操作等知识，希望大家通过本章的学习能够对Helm有一个初步的认识，能够在Operator开发过程中使用Helm。如果大家对Helm感兴趣，想要进一步深入学习Helm，可以仔细阅读其官方文档。